London Mathematical Society Lecture Note Series. 94

KU-240-802

Representations of General Linear Groups

G.D. JAMES

Fellow of Sidney Sussex College, Cambridge

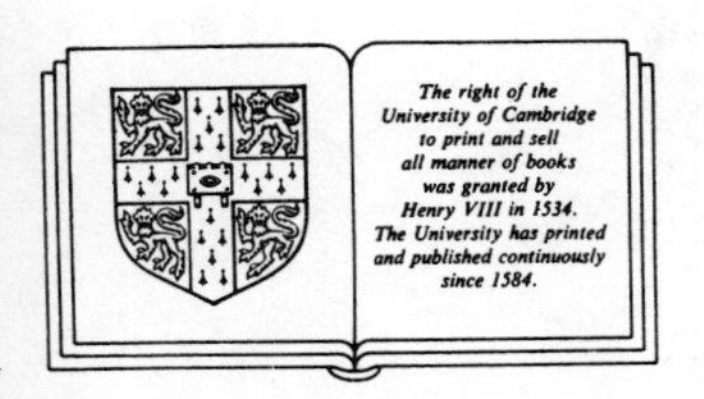
The right of the
University of Cambridge
to print and sell
all manner of books
was granted by
Henry VIII in 1534.
The University has printed
and published continuously
since 1584.

CAMBRIDGE UNIVERSITY PRESS

Cambridge

London New York New Rochelle

Melbourne Sydney

Published by the Press Syndicate of the University of Cambridge
The Pitt Building, Trumpington Street, Cambridge CB2 1RP
32 East 57th Street, New York, NY 10022, USA
296 Beaconsfield Parade, Middle Park, Melbourne 3206, Australia

First published 1984

Printed in Great Britain at the University Press, Cambridge

Library of Congress catalogue card number: 83-25171

British Library Cataloguing in Publication Data

James, G.D.
Representations of general linear groups -
(London Mathematical Society lecture note
series, ISSN 0076-0552; 94)
1. Linear algebraic groups
I. Title II. Series

ISBN 0 521 26981 4

Contents

ABSTRACT

This essay concerns the unipotent representations of the finite general linear groups $GL_n(q)$. An irreducible unipotent representation is, by definition, a composition factor of the permutation representation of $GL_n(q)$ on a Borel subgroup, and the ordinary irreducible unipotent representations may be indexed by partitions λ of n, as may the ordinary irreducible representations of the symmetric group $\mathfrak{S}_n$. The remarkable feature is that the representation theory of $\mathfrak{S}_n$ over an arbitrary field appears to be the case "q = 1" of the subject we study here.

The most important results are undoubtedly the Submodule Theorem (Chapter 11) and the Kernel Intersection Theorem (Chapter 15), but there seems to have been no previous work on the representation modules for the unipotent representations of $GL_n(q)$, so we claim originality for all the results apart from those whose source is quoted or which are obviously known (Chapters 3 - 8).

Chapters 1 and 2 set the scene, by outlining the connection between $\mathfrak{S}_n$ and representations of $GL_n(q)$ over fields of characteristic dividing q, and by giving examples of the situation to be considered later. The preliminary results which we need are derived in Chapters 3 - 8. Thereafter, we assume that the characteristic of our ground field K does not divide q, but otherwise K is arbitrary. Certain idempotents of the group algebra are defined in Chapter 9, and they are used in Chapter 10 to describe the structure of the permutation module M_λ of $GL_n(q)$ on a parabolic subgroup.

In Chapter 11, we define a certain submodule S_λ of M_λ in terms of a generator; S_λ may be regarded as the q-analogue of a Specht module. The Submodule Theorem states that every $KGL_n(q)$-submodule of M_λ either contains S_λ or is contained in $S_\lambda^\perp$. We proved the Submodule Theorem for $\mathfrak{S}_n$ in 1976 (James [J_1]), and thereby gave the first construction of the irreducible representations of $\mathfrak{S}_n$ over an arbitrary field. We have already published a proof of the Submodule Theorem for $GL_n(q)$ (James [J_9]), but the proof given here is new; it is simplified by assuming initially that the ground field contains all the pth roots of unity (where q is a power of p). The Submodul Theorem gives us an irreducible unipotent representation of $GL_n(q)$ for each partition of n. In particular, the various modules S_λ are the ordinary irreducible unipotent representations when the set of rational numbers is the ground field.

The aim of the next few chapters is to construct a basis for S_λ, and to prove the Kernel Intersection Theorem (Chapter 15), which describes S_λ as the intersection of the kernels of certain $KGL_n(q)$-homomorphisms defined on M_λ. Here we roughly follow the approach we adopted in 1977 (James [J_4]) to prove similar results for Specht modules. Unlike the situation for symmetric groups, where bases for Specht modules and the Kernel Intersection Theorem are easy for many special cases, the only partitions for which the $GL_n(q)$ results are clear are (n) (when there is nothing to prove!) and (n - 1, 1). Even the partition (2, 2) of 4 is difficult to handle; in place of a 2-dimensional representation of $\mathfrak{S}_4$, we have to deal with a $(q^2 + q^4)$-dimensional representation of $GL_4(q)$.

Many important results (Chapter 16) follow from the Kernel Intersecti Theorem. For example, dim S_λ is shown to be independent of K, and we prove that we have found all the irreducible unipotent representations over K. The Branching Theorem, describing the structure of S_λ as a $KGL_{n-1}(q)$-module, is also deduced.

By combining the Submodule Theorem and the Kernel Intersection Theore it is possible to embark upon the task of finding the decomposition matrices of $GL_n(q)$ for primes which do not divide q. The problem of determining the decomposition matrices of $\mathfrak{S}_n$ is still open, and we believe that the key may well lie with the unipotent representations of $GL_n(q)$.

In Chapter 17, we prove a theorem on the decomposition matrix of

$GL_n(q)$ concerning the removal of the first column from the diagram $[\lambda]$; the corresponding $\mathfrak{S}_n$ result was proved only recently (James $[J_8]$).

As far as we know, only the parts of the decomposition matrix of $\mathfrak{S}_n$ corresponding to hook partitions or to two-part partitions is known (Peel [P] and James $[J_2, J_3]$), although work is in progress on the partitions $(n - m - 1, m, 1)$. An analogue of Peel's results is given in Chapter 16, and in the final two chapters we determine the part of the decomposition matrix of $GL_n(q)$ which corresponds to two-part partitions, for all primes which do not divide q; the evidence that the modular representation theory of $\mathfrak{S}_n$ is just the case "q = 1" is then overwhelming.

Naturally, we have pondered the question why the modular representations of $\mathfrak{S}_n$ look like representations of the group of automorphisms of an n-dimensional vector space over "the field of one element". It is easy to be misled into giving an unsound argument about this, and it must be noted that our proofs do not translate into proofs for $\mathfrak{S}_n$. More challenging still is the explanation of the possible result that the representation theory over $\mathbb{F}_r$ of $GL_d(r)$ $(d \geq n$, r prime) is the case "q = 1" of our work here - see the remarks at the end of Chapter 16. Why should the representation theory of $GL_n(q)$ over fields whose characteristic does not divide q throw light on the representation theory of general linear groups of different dimension over fields of the natural characteristic?

Knowledge of the theory for $\mathfrak{S}_n$ has guided us to search for proofs to present here which would translate immediately into proofs for the symmetric group. We have been unsuccessful, so we cannot explain why "putting q = 1" works, and entirely new techniques have had to be developed in this essay.

LIST OF SYMBOLS

Symbol	Meaning	Chapter of definition
A_r	A certain subgroup of U^-	8.1
$A_r^{(i)}$	A certain idempotent of $\bar{K}A_r$	9.8
$B^{\pm}$	The group of upper/lower triangular matrices	5
c	A function from Γ to $\{1, 2, \ldots, q\}$	9.1
c_r	A certain function from $\Gamma(r)$ to $\{1, 2, \ldots, q\}$	9.4
E	An idempotent in $\bar{K}G_n$	
E_r	An idempotent in $\bar{K}G(\Gamma(r))$	9.4
E_λ	An idempotent in $\bar{K}U^-$	11.4
e	The exponent of q modulo p	19.4
$e_1, e_2, \ldots$	A basis for V	2.4
$\mathbb{F}_q$	The field of q elements	2.4
$(\mathbb{F}_q, +)$	The additive group of $\mathbb{F}_q$	
G_r	A subgroup of $GL_n(q)$, isomorphic to $GL_r(q)$	8.1
G_r^*	A certain subgroup containing G_r	8.1
$G(\Gamma)$	$\langle X_{ij} \mid (i, j) \in \Gamma \rangle$	5.1
$GL_n(q)$	The group of automorphisms of V	2.4
H	The group of diagonal matrices	5
H_r^*	A certain subgroup contained in G_r^*	8.1
h	The number of non-zero parts of λ	4
$h_i(\beta)$	A certain diagonal matrix	5
h_{ij}	The hook length of the (i, j) node in $[\lambda]$	20.1
I_r	The identity $r \times r$ matrix	
K	A field of characteristic coprime to q	
$\bar{K}$	K extended by a primitive p^{th} root of unity, where q is a power of p	8
$\ell_p(m)$	The least non-negative integer i such that $m < p^i$	19.1

M_λ	The permutation module on P_λ	10.1
$[m]$	$1 + q + q^2 + \ldots + q^{m-1}$	2.5
$\{m\}$	$[1]\ [2]\ \ldots\ [m]$	10.17
n	The dimension of V	2.4
$\begin{bmatrix} n \\ m \end{bmatrix}$	A Gaussian polynomial	2.14
$\begin{Bmatrix} n \\ m \end{Bmatrix}$	A binomial coefficient	
P_λ	A parabolic subgroup	6
p	A prime number	
$\mathbb{Q}$	The field of rational numbers	
q	A power of a prime number	
R	A subset of $\{1, 2, \ldots, h\}$	10.8
$\mathcal{R}_r$	The set of subsets of $\{1, 2, \ldots, h\}$ of cardinality r	10.8
$\mathcal{R}_r^*$	A certain subset of $\mathcal{R}_r$	10.20
S_λ	A certain submodule of M_λ	11.11
$\mathfrak{S}_n$	The symmetric group on n symbols	
T_λ	The initial λ-tableau	4.2
$U^\pm$	The group of upper/lower unitriangular matrices	5.6
$U_\lambda^\pm$	A certain subgroup of $U^\pm$	6.1
V	The n-dimensional vector space over $\mathbb{F}_q$ of which $GL_n(q)$ is the group of automorphisms	2.4
W	The group of permutation matrices	5
X_{ij}	A root subgroup	5
$x_{ij}(\alpha)$	An element of X_{ij}	5
$\mathbb{Z}$	The ring of integers	
$\alpha, \beta, \gamma, \delta$	Elements of $\mathbb{F}_q$	
Γ	A closed subset of Φ	5.1
Γ'	The "commutator" subset of Γ	5
$\Gamma(r)$	$\{(i, j) \mid n \geqq i > j \leq r \leq n\}$	9.4
θ	A $\bar{K}G_n$-homomorphism	

κ	An element of K	
λ, μ, ν	Compositions of n	4
$\nu_p(m)$	The largest integer i such that p^i divides m	19.1
π, σ, τ	Permutations	
π_λ	A certain permutation, depending on λ	11.3
π_R	A certain permutation, depending on R	12.1
Φ	$\{(i, j) \mid 1 \leqq i \neq j \leqq n\}$	5
Φ^+	$\{(i, j) \mid 1 \leqq i < j \leqq n\}$	5.6
Φ^-	$\{(i, j) \mid 1 \leqq j < i \leqq n\}$	5.6
$\phi_1, \phi_2, \ldots$	Linear $\bar{K}$-characters of A_r	9.6
$\chi_1, \chi_2, \ldots$	Linear $\bar{K}$-characters of $(\mathbb{F}_q, +)$	9.1
χ_c	A linear $\bar{K}$-character of $G(\Gamma)$	9.1
χ_λ	The ordinary character of S_λ	1
$\psi_{d,i}$	A certain $\bar{K}G_n$-homomorphism defined on M_λ	15.1
$\trianglerighteq$	A transitive relation on the set of compositions of n	4.1
$\triangleright$	$\trianglerighteq$ but not =	4.1
$\langle\ldots\rangle$	The group, or vector space, generated by ...	
$\langle\ ,\ \rangle_\lambda$	A bilinear form on M_λ	11.1

1 INTRODUCTION

Among all classes of groups, it is arguable that the symmetric groups $\mathfrak{S}_n$ have the richest representation theory. Not only are the symmetric groups interesting in their own right - the theory of their representations is extremely elegant, and still contains many mysteries - but they can also be used in several ways to shed light on the representations of other groups, and their theory can be applied in fields as diverse as quantum mechanics and polynomial identity algebras. We hope to convince the reader that the representation theory of symmetric groups is just a special case of a far deeper, but equally interesting, topic, namely the theory of unipotent representations of the finite general linear groups $GL_n(q)$.

It is well-known that there is a close connection between representations of $\mathfrak{S}_n$ over a field F and the representations of $GL_n(F)$ over the same field F. But this is not what we shall explore here; instead, we open up a new avenue by considering representations of $GL_n(q)$ over a field K whose characteristic <u>does not divide q</u>.

The ordinary irreducible characters of $GL_n(q)$ have been determined by Green [G_1], but earlier work of Steinberg [S] produced one ordinary irreducible character χ_λ for each partition λ of n. It appears that no previous work has been done on constructing the representation modules for $GL_n(q)$, and we shall deal here entirely with the modules corresponding to the characters obtained by Steinberg, the so-called <u>unipotent representations</u> of $GL_n(q)$. We shall be working, then, with representations of $GL_n(q)$ which are indexed by partitions λ of n. The ordinary irreducible representations of $\mathfrak{S}_n$ are also indexed this way. Two striking results about the characters χ_λ of $GL_n(q)$ have been proved which already indicate an analogy with $\mathfrak{S}_n$:

1.1 THEOREM (Olsson [O]). If $\lambda = (\lambda_1, \lambda_2, \ldots)$ is a partition of n then for the general linear group

$$\deg \chi_\lambda = q^{\sum_k (k-1)\lambda_k} (q^n - 1)(q^{n-1} - 1) \cdots (q-1) \prod_h \frac{1}{(q^h - 1)},$$

where the product is over the hook lengths $h = (\lambda_i + \lambda'_j + 1 - i - j)$ in the the diagram $[\lambda]$.

(Olsson gives a similar formula for all the character degrees of $GL_n(q)$.)

1.2 THEOREM (Fong and Srinivasan [FS]). Assume that p is an odd prime not dividing q, and e is the least positive integer such that p divides $q^e - 1$. Then χ_λ and χ_μ are in the same p-block of $GL_n(q)$ if and only if $[\lambda]$ and $[\mu]$ have the same e-core.

Compare these results with two theorems from the representation theory of $\mathfrak{S}_n$ (We shall also denote by χ_λ the ordinary irreducible character of $\mathfrak{S}_n$ corresponding to the partition λ of n.)

1.3 THEOREM (Frame, Robinson and Thrall [FRT]). For the symmetric group

$$\deg \chi_\lambda = n(n-1) \ldots 1 \prod_h \left(\frac{1}{h}\right).$$

1.4 THEOREM (Brauer [B] and Robinson [R]). χ_λ and χ_μ belong to the same p-block of $\mathfrak{S}_n$ if and only if $[\lambda]$ and $[\mu]$ have the same p-core

What is more, there are similar results in the theory of Weyl modules. We shall not use Weyl modules here, so we shall not give a formal definition of them, but refer the reader to the relevant literature (for example, Green [G_2] or James and Kerber [JK]). Suffice it to say that if F is a sufficiently large field, and V is the d-dimensional vector space over F on which $GL_d(F)$ acts in the natural way, then for each non-negative integer n and for every partition λ of n having at most d non-zero parts, there is a $GL_d(F)$-submodule W_λ, called a Weyl module, of $V^{\otimes n}$. We emphasize that W_λ is a representation module for $GL_d(F)$ over the "natural" field F.

1.5 THEOREM <u>The dimension of W_λ is independent of F, and</u>

$$\dim W_\lambda = \prod_{(i,j)\in[\lambda]} (d + j - i) \prod_h \left(\frac{1}{h}\right) .$$

Two Weyl modules W_λ and W_μ are said to be connected if there exists a sequence $\lambda = \lambda_1, \lambda_2, \ldots , \lambda_k = \mu$ of partitions such that for each i, W_{λ_i} and $W_{\lambda_{i+1}}$ have a common composition factor.

1.6 THEOREM <u>If F has characteristic p and λ, μ are partitions of the same integer, then W_λ and W_μ are connected if and only if $[\lambda]$ and $[\mu]$ have the same p-core.</u>

In fact, the theory of Weyl modules is much closer to the symmetric group than one might expect. For each λ, W_λ has a unique maximal submodule; we denote the quotient by F_λ, whereupon every irreducible polynomial representation of $GL_d(F)$ is isomorphic to some F_λ. For each partition λ of n, there is a Specht module, defined over F, for $\mathfrak{S}_n$. Most Specht modules have a unique maximal submodule; the various quotients give all the irreducible $F\mathfrak{S}_n$-modules. To illustrate the connection between $F\mathfrak{S}_n$-modules and $FGL_d(F)$-modules, and also for future reference, we now give an example.

1.7 EXAMPLE. The following matrices describe the composition factors of some Weyl modules. The entry in row λ and column μ is the number of composition factors of W_λ which are isomorphic to F_μ.

$$\begin{array}{c} \\ (1,1) \\ (2) \end{array} \begin{array}{c} \begin{array}{cc} (1,1) & (2) \end{array} \\ \left[\begin{array}{cc} \underline{1} & 0 \\ \underline{1} & 1 \end{array}\right] \end{array} \quad \text{if char}\, F = 2$$

$$\begin{array}{r} \\ (1,1,1) \\ (2,1) \\ (3) \end{array} \begin{array}{ccc} (1,1,1) & (2,1) & (3) \\ \hline \end{array}$$

$$\begin{array}{r|ccc|} & (1,1,1) & (2,1) & (3) \\ \hline (1,1,1) & \underline{1} & \underline{0} & 0 \\ (2,1) & \underline{0} & \underline{1} & 0 \\ (3) & \underline{1} & \underline{0} & 1 \end{array} \quad \text{if char } F = 2$$

$$\begin{array}{r|ccc|} & (1,1,1) & (2,1) & (3) \\ \hline (1,1,1) & \underline{1} & \underline{0} & 0 \\ (2,1) & \underline{1} & \underline{1} & 0 \\ (3) & \underline{0} & \underline{1} & 1 \end{array} \quad \text{if char } F = 3$$

$$\begin{array}{r|ccccc|} & (1^4) & (2,1^2) & (2,2) & (3,1) & (4) \\ \hline (1^4) & \underline{1} & \underline{0} & 0 & 0 & 0 \\ (2,1^2) & \underline{1} & \underline{1} & 0 & 0 & 0 \\ (2,2) & \underline{0} & \underline{1} & 1 & 0 & 0 \\ (3,1) & \underline{1} & \underline{1} & 1 & 1 & 0 \\ (4) & \underline{1} & \underline{0} & 1 & 1 & 1 \end{array} \quad \text{if char } F = 2$$

$$\begin{array}{r|ccccc|} & (1^4) & (2,1^2) & (2,2) & (3,1) & (4) \\ \hline (1^4) & \underline{1} & \underline{0} & \underline{0} & \underline{0} & 0 \\ (2,1^2) & \underline{0} & \underline{1} & \underline{0} & \underline{0} & 0 \\ (2,2) & \underline{1} & \underline{0} & \underline{1} & \underline{0} & 0 \\ (3,1) & \underline{0} & \underline{0} & \underline{0} & \underline{1} & 0 \\ (4) & \underline{0} & \underline{0} & \underline{1} & \underline{0} & 1 \end{array} \quad \text{if char } F = 3$$

Remember that all these matrices give information about every general linear group $GL_d(F)$ over F. The partitions are not partitions of the integer d. On the other hand, the matrix involving partitions of n contains the decomposition matrix of $\mathfrak{S}_n$ over F (James [J_7]); the underlined entries give this decomposition matrix. It can also be proved (James [J_{10}]) that every column of the matrix for n corresponds in a natural way to an indecomposable module for $F\mathfrak{S}_n$.

The results which are illustrated in the example above are true in general. If we had complete information about the components of permutation modules of symmetric groups, then we could construct all the matrices which describe the composition factors of Weyl modules, and conversely, since these matrices contain the decomposition matrices of symmetric groups, knowledge of the composition factors of Weyl modules would give the decomposition matrices of symmetric groups.

We hope now to have given sufficient support to the claim that the representation theory of $\mathfrak{S}_n$ is very closely tied to that of $GL_n(q)$ over fields whose characteristic divides q. How then does the representation theory of $GL_n(q)$ over fields of characteristic <u>not</u> dividing q fit into the picture? We aim to show that the representation theory of $\mathfrak{S}_n$ must be "the case $q = 1$" of this theory. It is well-known that $\mathfrak{S}_n$ looks like "the general linear group over the field of one element", but the results go through in an unexpected and beautiful way. We emphasise that the methods used in this essay do not apply to the symmetric groups; everything we achieve is <u>analogous</u> to the theory of symmetric groups, but whenever we suggest that q should be put equal to 1, we do so for the results but not for the proofs. We do not know why this works.

If K is a field whose characteristic does not divide q, we shall produce an irreducible $KGL_n(q)$-module for each partition of n. In general, there are fewer irreducible $K\mathfrak{S}_n$-modules than there are partitions of n, but there is a Weyl module for each partition of n. Could our new theory subsume the theory of Weyl modules, too? The evidence is flimsy, but consistent (see Chapter 16).

A light-hearted observation concerns the tendency of theorems from the representation theory of symmetric groups to take a form which does not rely on the primeness of the field characteristic (for example, several theorems involve the p-core of a diagram, which exists whether or not p is prime.) It appears that this phenomenon occurs because, for example, there are primes which divide $1 + q + q^2 + q^3$ but not $1 + q$; when we put $q = 1$ the general linear group results for these primes turn into empty theorems about those primes dividing 4 and not 2.

It will be clear by the end of this essay that there are still many interesting open questions. A recent volume on the representation theory of $\mathfrak{S}_n$ by the present author and Adalbert Kerber [JK] contained nigh on a thousand references to works on that subject. We believe that this is just the tip (q = 1) of a very big iceberg!

2 EXAMPLES

Few of the results from this section will be used later on, but it is enlightening to look at some special cases of the problems we shall consider, before plunging into the general situation.

First we consider a representation of $GL_n(q)$, denoted by $(n - 2,2)$, which looks relatively straightforward. It is already difficult, though, to complete the relevant calculations for this representation.

The symmetric group $\mathfrak{S}_n$ is the group of all permutations of $\{1, 2, \ldots, n\}$. We define a representation of $\mathfrak{S}_n$ which is indexed by $(n - 2,2)$, and then show what happens for the general linear group.

Consider a vector space, over a field $\mathbf{K}$, whose basis elements are the unordered pairs $\{i, j\}$ from $\{1, 2, \ldots, n\}$. Thus, for example,

$$\kappa_1\{1, 2\} + \kappa_2\{3, 4\} \qquad (\kappa_1, \kappa_2 \in \mathbf{K})$$

belongs to our vector space. Since $\mathfrak{S}_n$ permutes the pairs $\{i, j\}$, our vector space may be regarded as a $\mathbf{K}\mathfrak{S}_n$-module. The dimension of the space is simply the number of unordered pairs,

$$\frac{n(n-1)}{2}.$$

Let $S = S_{(n-2,2)}$ be the subspace consisting of those vectors satisfying the following two conditions:

(0) The sum of the coefficients is zero. (Thus we require that $\kappa_1 + \kappa_2 = 0$ in our example above, for this condition to hold.)

(1) For each 1-element subset $\{u\}$ of $\{1, 2, \ldots, n\}$, the sum of the coefficients of the pairs containing u is zero.

The subspace S is a $\mathbf{K}\mathfrak{S}_n$-module, since conditions (0) and (1) are preserved under the action of $\mathfrak{S}_n$. An example of a vector satisfying conditions (0) and (1) is:

(2.1) $\{12\} - \{23\} + \{34\} - \{41\}$,

for certainly the sum of the coefficients $(+1 - 1 + 1 - 1)$ is zero, and if we look at $u = 2$, for example, the pairs containing u are $\{12\}$ and $\{23\}$, and the sum of the coefficients $(+1 - 1)$ of these pairs is zero. That conditions (0) and (1) are fulfilled is seen most easily by drawing a picture to describe the vector (2.1):

(2.2)

1 +1 2

-1 -1

4 +1 3

The following are facts about the space S:

(2.3) (i) S is non-zero if and only if $n \geq 4$.

(ii) S has a basis consisting of vectors of the form (2.2).

(iii) Dim $S = n(n - 3)/2$.

(iv) If K has characteristic zero, then S is an irreducible $K\mathfrak{S}_n$-modu

The first three of these results are true regardless of the field K

We now turn to the general linear group, and choose some notation which will remain in force throughout this work:

2.4 DEFINITIONS. Let $\mathbb{F}_q$ be the field of q elements, and let V be a vector space over $\mathbb{F}_q$, with basis $e_1, e_2, \ldots, e_n$. We denote the general linear group $GL_n(q)$ by G_n. By definition, G_n is the group of all automorphisms of V. We shall freely identify G_n with the group of non-singular $n \times n$ matrices over $\mathbb{F}_q$, the automorphism given by the matrix (g_{ij}) being the one for which

$$e_i \mapsto \sum_{j=1}^{n} g_{ij} e_j \qquad (1 \leq i \leq n) .$$

If $v_1, v_2, \ldots, v_k$ are vectors in V, we let

$\langle v_1, v_2, \ldots, v_k \rangle$

denote the subspace of V spanned by $v_1, v_2, \ldots, v_k$.

2.5 DEFINITION. If m is a non-negative integer, let

$$[m] = 1 + q + q^2 + \ldots + q^{m-1} .$$

In particular, $[0] = 0$, $[1] = 1$, and $[m] = (q^m - 1)/(q - 1)$.

Note that if we put $q = 1$ in this definition, we get $[m] = m$. Bear this in mind while we go through a "q-analogue" of our statements about $\mathfrak{S}_n$.

Consider a vector space over K whose basis elements are the 2-dimensional subspaces of V. Thus, for example,

$$\kappa_1\langle e_1, e_2\rangle + \kappa_2\langle e_3, e_4\rangle \qquad (\kappa_1, \kappa_2 \in K)$$

belongs to our vector space. Since G_n permutes the 2-dimensional subspaces of V, our vector space may be viewed as a KG_n-module. The dimension of this space is simply the number of 2-dimensional subspaces of V, which, by a simple calculation (see Theorem 3.1) is

$$\frac{[n][n-1]}{[2]} .$$

Consider the subspace $S = S_{(n-2,2)}$ consisting of those vectors satisfying the following two conditions:

(2.6) (0) The sum of the coefficients is zero. (Thus, we require that $\kappa_1 + \kappa_2 = 0$ in our example above, for this condition to hold.)

(1) For each 1-dimensional subspace U of V, the sum of the coefficients of the 2-dimensional subspaces containing U is zero.

The subspace S is a KG_n-module, since conditions (0) and (1) are preserved under the action of G_n.

We recommend at this stage that the reader attempts for himself to construct a non-zero element of S. It is conceivable (and after some effort, the reader might believe likely!) that S is the zero subspace; this would not be very interesting.

To exhibit a non-zero element of S, we resort to the notation of projective geometry. A line will denote a 2-dimensional subspace U of V, and the points on the line will be the 1-dimensional subspaces of U.

EXAMPLE. If $q = 2$, $\langle e_1, e_2 \rangle$ contains three 1-dimensional subspaces $\langle e_1 \rangle$, $\langle e_2 \rangle$, and $\langle e_1 + e_2 \rangle$. In the picture below, the line through the points $\langle e_1 \rangle$, $\langle e_2 \rangle$, $\langle e_1 + e_2 \rangle$ is labelled +1; this denotes the fact that in our linear combination of 2-dimensional subspaces, $\langle e_1, e_2 \rangle$ occurs with coefficient +1.

(2.7)

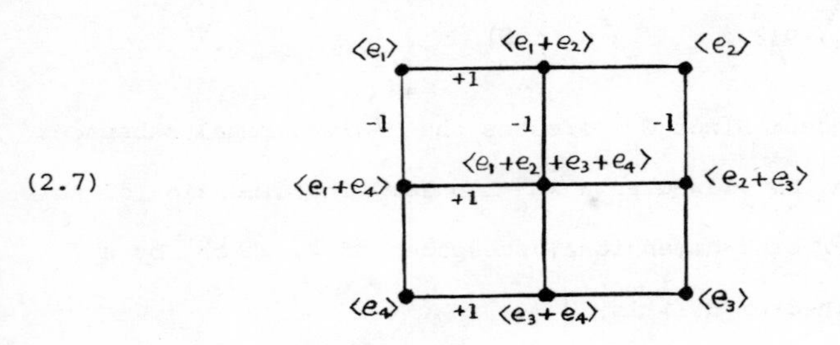

The picture corresponds to

$$\langle e_1, e_2 \rangle + \langle e_1 + e_4, e_2 + e_3 \rangle + \langle e_3, e_4 \rangle$$
$$- \langle e_1, e_4 \rangle - \langle e_1 + e_2, e_3 + e_4 \rangle - \langle e_2, e_3 \rangle ,$$

an element of the vector space over K whose basis elements are the 2-dimensional subspaces of V. A glance at the picture shows that conditions 2.6 hold, so we have constructed an element of S.

EXAMPLE. If $q = 3$, $\langle e_1, e_2 \rangle$ contains four 1-dimensional subspaces, $\langle e_1 \rangle$, $\langle e_2 \rangle$, $\langle e_1 + e_2 \rangle$, $\langle e_1 - e_2 \rangle$.

(2.8)

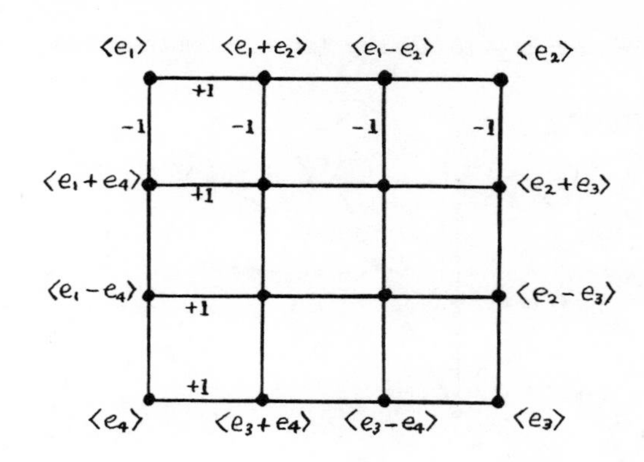

We leave the reader to label the four points in the middle of the picture; for example, the top left-hand point is $\langle e_1 + e_2 + e_3 + e_4 \rangle$, being the 1-dimensional intersection of $\langle e_1 + e_4, e_2 + e_3 \rangle$ and $\langle e_1 + e_2, e_3 + e_4 \rangle$. Again, we have an element of S.

Compare (2.8; q = 3), and (2.7; q = 2), with (2.2; q = 1?).

Notice that both the above Examples used four linearly independent vectors e_1, e_2, e_3, e_4. In fact, all the following statements are true for S:

(2.9) Assume that the characteristic of K does not divide q. Then

(i) S is non-zero if and only if $\dim V \geq 4$.

(ii) S has a basis consisting of vectors of the type illustrated by (2.7) and (2.8).

(iii) $\dim S = q^2[n][n-3]/[2]$.

(iv) If K has characteristic zero, then S is an irreducible KG_n-module.

Compare these statements with (2.3). The formula for dim S may not be the expected one, but still, if we put q = 1, then ... .

The first three of the statements above are hard to prove from the definition of S. One of the reasons for the difficulty is that the results are false - unless we use the hypothesis that the characteristic of K does not divide q. Let us show how things go wrong when we omit this condition.

2.10 EXAMPLE. Consider the following linear combinations of 2-dimensional subspaces of V:

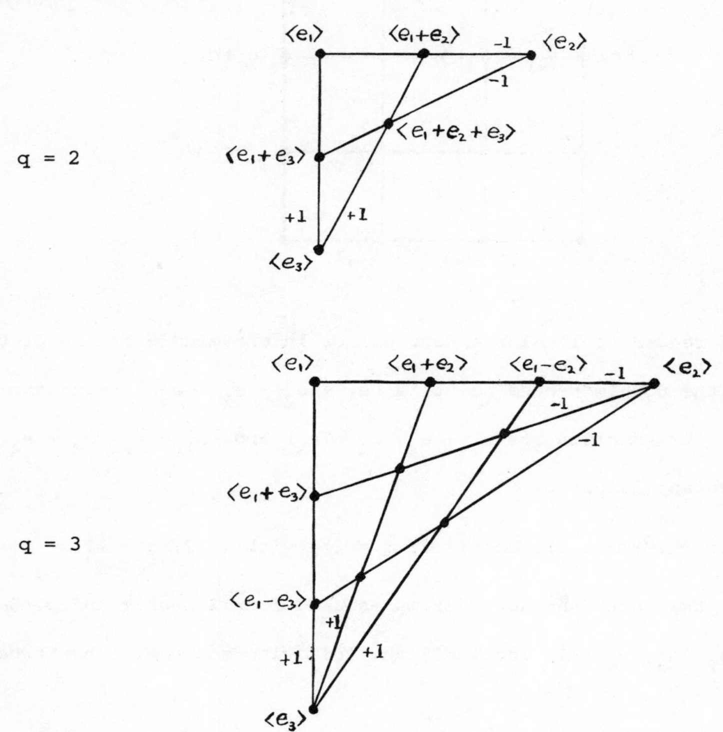

(The remaining points are labelled $\langle e_1 + e_2 + e_3\rangle$, $\langle e_1 - e_2 + e_3\rangle$, $\langle e_1 - e_2 - e_3\rangle$, $\langle e_1 + e_2 - e_3\rangle$, reading clockwise from the top left.)

These vectors satisfy condition 2.6(0), and for all 1-dimensional subspaces U of V, except $U = \langle e_2\rangle$ or $U = \langle e_3\rangle$, the sum of the coefficients of the 2-dimensional subspaces containing U is zero. But if we let our coefficient field K have characteristic q, then condition 2.6(1) holds for $U = \langle e_2\rangle$ and $U = \langle e_3\rangle$, also. Thus, if we allow the characteristic of K to divide q, we can construct a non-zero element of S when dim V = 3, contradict 2.9 (i) - (iii). ■

So far, we have concentrated on the 2-dimensional subspaces of V. But there is an obvious way of generalising the ideas to other cases. Let u see what happens.

Define $M_{(n-m,m)}$ to be a vector space over K whose basis elements are the m-dimensional subspaces of V. Thus, $M_{(n-m,m)}$ is a "coordinatized" version of the set of m-dimensional subspaces of V. Formally, we could regard $M_{(n-m,m)}$ as the vector space of K-valued functions on the set of m-dimensional subspaces of V, but we find our description conceptually easier. In fact, we shall later perform all our calculations with a right ideal of the group algebra KG_n which is isomorphic to $M_{(n-m,m)}$. The plan is to construct inside $M_{(n-m,m)}$ a KG_n-module $S_{(n-m,m)}$ which is the q-analogue of the Specht module for $\mathfrak{S}_n$ corresponding to (n-m,m). In Chapter 11, we shall define a module S_λ for each partition λ of n.

To provide examples of what we are about, we need a notation for the m-dimensional subspaces of V. Although we were able to draw lines for the 2-dimensional subspaces, something more elaborate is required for a general m.

An element of V may be regarded as a n-tuple (a row vector of length n) of elements from $\mathbb{F}_q$. Explicitly,

$$\alpha_1 e_1 + \alpha_2 e_2 + \dots + \alpha_n e_n \in V \qquad (\alpha_i \in \mathbb{F}_q)$$

corresponds to the row vector

$$(\alpha_1, \alpha_2, \dots, \alpha_n) .$$

Given m elements of V, viewed this way, we construct an $m \times n$ matrix by placing the row vectors one after the other. If the m elements of V are linearly independent, then the matrix has rank m, and corresponds to an m-dimensional subspace of V. This subspace is unchanged by applying each of the following operations to the matrix:

(2.11) (i) Divide all entries in a row by a non-zero element of $\mathbb{F}_q$.

(ii) Subtract a multiple of one row from another.

(iii) Rearrange the order of the rows.

By applying (i) and (ii) we can arrange that the last non-zero entry in each row is a 1 and that the other entries in the column containing this 1 are zero. Suppose, having done this, that the last non-zero entry in row i is a 1 in column a_i. Using 2.11(iii), we may assume that $1 \le a_1 < a_2 < \ldots < a_m \le n$. Thus, we can get our matrix in the form:

$$(2.12)\qquad \underbrace{\left.\begin{bmatrix} ** & \cdots & *\overset{a_1}{1}00 & \cdots & \overset{a_2}{0}00 & \cdots & \overset{a_3}{0}.. & .. & .. & 0 \\ ** & \cdots & *0** & \cdots & 100 & \cdots & 0.. & .. & .. & 0 \\ ** & \cdots & *0** & \cdots & 0** & \cdots & 1.. & .. & .. & 0 \\ & & & & & & & & & \\ \vdots & & \vdots & & \vdots & & \vdots & & & \end{bmatrix}\right\} m}_{n}$$

It is clear that there is a bijection between the set of matrices in this "canonical form" and the set of m-dimensional suspaces of V.

2.13 EXAMPLE. $n = 4$, $m = 2$. Each 2-dimensional subspace of V correspon to one of the following matrices:

$$\begin{pmatrix} 1 & 0 & 0 & 0 \\ 0 & 1 & 0 & 0 \end{pmatrix} \quad \begin{pmatrix} 1 & 0 & 0 & 0 \\ 0 & \alpha & 1 & 0 \end{pmatrix} \quad \begin{pmatrix} 1 & 0 & 0 & 0 \\ 0 & \alpha & \beta & 1 \end{pmatrix} \quad \begin{pmatrix} \alpha & 1 & 0 & 0 \\ \beta & 0 & 1 & 0 \end{pmatrix}$$

$$\begin{pmatrix} \alpha & 1 & 0 & 0 \\ \beta & 0 & \gamma & 1 \end{pmatrix} \quad \begin{pmatrix} \alpha & \beta & 1 & 0 \\ \gamma & \delta & 0 & 1 \end{pmatrix}$$

($\alpha, \beta, \gamma, \delta \in \mathbb{F}_q$). Thus there are $1 + q + 2q^2 + q^3 + q^4$ 2-dimensional subspa of V. ■

2.14 DEFINITION (see also 2.5).

Let $\begin{bmatrix} n \\ m \end{bmatrix} = \dfrac{[n][n-1]\ \ldots\ [n-m+1]}{[m][m-1]\ \ldots\ [1]}$, if $n \ge m \ge 0$

$= 0$, otherwise.

The number $\begin{bmatrix} n \\ m \end{bmatrix}$ is a q-analogue of the binomial coefficient $\begin{pmatrix} n \\ m \end{pmatrix}$; when we put $q = 1$, we get

$$\begin{bmatrix} n \\ m \end{bmatrix} = \binom{n}{m} .$$

t is easy to show (cf. Theorem 3.1) that the number of m-dimensional subspaces f V is $\begin{bmatrix} n \\ m \end{bmatrix}$.

XAMPLE.

$$\begin{bmatrix} 4 \\ 2 \end{bmatrix} = \frac{[4][3]}{[2][1]} = \frac{(1 + q + q^2 + q^3)(1 + q + q^2)}{(1 + q)}$$

$$= (1 + q^2)(1 + q + q^2) = 1 + q + 2q^2 + q^3 + q^4 .$$

rom Example 2.13, we see that this is, indeed, the number of 2-dimensional ubspaces of a 4-dimensional space. ∎

Note that the arrangement of *'s in (2.12) corresponds to a partition, amely

$$(a_m - m, \ldots, a_2 - 2, a_1 - 1) .$$

his partition has at most m non-zero parts, and each part is at most n - m. ince there is a bijection between the set of such partitions and the set of equences

$$1 \le a_1 < a_2 < \ldots < a_m \le n ,$$

he two different ways of counting the m-dimensional subspaces of V give:

.15 THEOREM (cf. Andrews [A] p. 33)

$$\begin{bmatrix} n \\ m \end{bmatrix} = \sum_{i=0}^{m(n-m)} a_{n,m}(i) q^i ,$$

here $a_{n,m}(i)$ equals the number of partitions of i, each of which has at most non-zero parts and largest part at most n - m. ∎

In particular, $\begin{bmatrix} n \\ m \end{bmatrix}$ is a polynomial in q; it is called a Gaussian olynomial.

2.16 DEFINITION. Assume that $0 \le i \le m$. Define $\psi_{1,i}$ to be the linear ma

from $M_{(n-m,m)}$ into $M_{(n-i,i)}$ which sends each m-dimensional subspace of V to

the sum of the i-dimensional subspaces contained in it.

The reason for the first subscript 1 in $\psi_{1,i}$ will emerge when we com

to define more general maps. It is clear that $\psi_{1,i}$ is a KG_n-homomorphism.

2.17 EXAMPLES

(i) $m = 2,\ i = 1$.

$$\psi_{1,i} : \langle e_1, e_2 \rangle \mapsto \langle e_1 \rangle + \sum_{\alpha \in \mathbb{F}_q} \langle \alpha e_1 + e_2 \rangle .$$

(ii) $m = 3$

$$\psi_{1,0} : \langle e_1, e_2, e_3 \rangle \mapsto \langle 0 \rangle .$$

$$\psi_{1,1} : \langle e_1, e_2, e_3 \rangle \mapsto \langle e_1 \rangle + \sum_{\alpha \in \mathbb{F}_q} \langle \alpha e_1 + e_2 \rangle + \sum_{\alpha,\beta \in \mathbb{F}_q} \langle \alpha e_1 + \beta e_2 +$$

$$\psi_{1,2} : \langle e_1, e_2, e_3 \rangle \mapsto \begin{pmatrix} 1 & 0 & 0 & 0 \\ 0 & 1 & 0 & 0 \end{pmatrix} + \sum_{\alpha \in \mathbb{F}_q} \begin{pmatrix} 1 & 0 & 0 & 0 \\ 0 & \alpha & 1 & 0 \end{pmatrix} + \sum_{\alpha,\beta \in \mathbb{F}_q} \begin{pmatrix} \alpha & 1 & 0 & 0 \\ \beta & 0 & 1 & 0 \end{pmatrix}$$

In the case where $m = 2$, a vector in $M_{(n-2,2)}$ belongs to $\operatorname{Ker} \psi_{1,0}$ if

and only if it satisfies Condition 2.6(0), and it belongs to $\operatorname{Ker} \psi_{1,1}$ if and

only if it satisfies Conditon 2.6(1).

In general, the subspace of $M_{(n-m,m)}$ on which we shall focus our

attention is

$$S_{(n-m,m)} = \bigcap_{i=0}^{m-1} \operatorname{Ker} \psi_{1,i} .$$

We must warn that we shall not define $S_{(n-m,m)}$ this way, and it is only aft

some hard work that we can prove that our module is a kernel intersection,

as above.

The reader who wishes to explore some of the intricacies of the subj

for himself is urged to construct a non-zero element of $S_{(n-m,m)}$ when

$= m \le n - m$, and to prove that $S_{(n-m,m)}$ is zero if $m > n - m$ and the haracteristic of K does not divide q. As a hint to the generator of $S_{(n-m,m)}$ hich we shall define later, draw the lines corresponding to

$$\sum_{\alpha \in \mathbb{F}_q} \left\{ \begin{pmatrix} 0 & 1 & 0 & 0 \\ \alpha & 0 & 0 & 1 \end{pmatrix} - \begin{pmatrix} 1 & 1 & 0 & 0 \\ \alpha & 0 & 0 & 1 \end{pmatrix} \right\},$$

nd compare the answer with Example 2.10. Then consider

$$\sum_{\alpha \in \mathbb{F}_q} \left\{ \begin{pmatrix} 0 & 1 & 0 & 0 \\ \alpha & 0 & 0 & 1 \end{pmatrix} - \begin{pmatrix} 1 & 1 & 0 & 0 \\ \alpha & 0 & 0 & 1 \end{pmatrix} - \begin{pmatrix} 0 & 1 & 0 & 0 \\ \alpha & 0 & 1 & 1 \end{pmatrix} + \begin{pmatrix} 1 & 1 & 0 & 0 \\ \alpha & 0 & 1 & 1 \end{pmatrix} \right\}.$$

or a generalisation of this idea for m arbitrary, see Example 11.17(v); to eneralize (2.7) and (2.8), see Chapter 18.

3 GAUSSIAN POLYNOMIALS

The Gaussian polynomials, which we introduced in the last section, behave in many ways like binomial coefficients. We wish to record some of their properties, referring to Andrews [A] for a fuller account.

3.1 THEOREM. *Let V_1, V_2 be subspaces of V with $\dim V_1 = a$, $\dim V_2 = b$, and $V_1 \cap V_2 = 0$. The number of m-dimensional subspaces W of V such that $W \cap V_1 = 0$ and $W \supseteq V_2$ is*

$$q^{a(m-b)} \begin{bmatrix} n-a-b \\ m-b \end{bmatrix} .$$

Proof: Given a basis of V_2, the number of ways of choosing a sequence of m - b further linearly independent vectors in such a way that no non-zero linear combination of them lies in $V_1 \oplus V_2$ is

$$(q^n - q^{a+b})(q^n - q^{a+b+1}) \ldots (q^n - q^{a+m-1})$$

$$= (q-1)^{m-b} q^{(m-b)(m+b+2a-1)/2} [n-a-b][n-a-b-1] \ldots [n-a$$

But an m-dimensional subspace containing V_2 has the following number of b extending the given basis of V_2:

$$(q^m - q^b)(q^m - q^{b+1}) \ldots (q^m - q^{m-1})$$

$$= (q-1)^{m-b} q^{(m-b)(m+b-1)/2} [m-b][m-b-1] \ldots [1] .$$

The quotient of these two numbers is the required answer.

Taking $V_1 = V_2 = 0$, we deduce that the number of m-dimensional subs of V is $\begin{bmatrix} n \\ m \end{bmatrix}$.

3.2 COROLLARY. *Let U, V_2 be subspaces of V with $V_2 \subseteq U$, $\dim V_2 = b$, $\dim U = c$. Then the number of m-dimensional subspaces W of V such that $W \cap U = V_2$ is*

$$q^{(c-b)(m-b)} \begin{bmatrix} n-c \\ m-b \end{bmatrix} .$$

Proof: Let V_1 be a complementary subspace to V_2 in U, and apply the Theorem. ■

3.3 COROLLARY. Let U be a subspace of V with dim U = c. Then

(i) The number of m-dimensional subspaces W of V such that $W \supseteq U$ is

$$\begin{bmatrix} n-c \\ m-c \end{bmatrix} .$$

(ii) The number of m-dimensional subspaces W of V such that $W \cap U = 0$ is

$$q^{cm} \begin{bmatrix} n-c \\ m \end{bmatrix} .$$

Proof: For part (i) put $V_2 = U$, and for part (ii) put $V_2 = 0$ in Corollary 3.2. ■

Either by direct calculation, or using Corollary 3.3 to count the number of m-dimensional subspaces of V which do or do not contain $\langle e_1 \rangle$, we get

3.4 THEOREM

$$\begin{bmatrix} n \\ m \end{bmatrix} = \begin{bmatrix} n-1 \\ m-1 \end{bmatrix} + q^m \begin{bmatrix} n-1 \\ m \end{bmatrix}$$
$$= \begin{bmatrix} n-1 \\ m \end{bmatrix} + q^{n-m} \begin{bmatrix} n-1 \\ m-1 \end{bmatrix} .$$

Applying this result, and using induction on n, one easily proves a q-analogue of the Binomial Theorem:

3.5 THEOREM. For all x,

$$(1+x)(1+xq) \dots (1+xq^{n-1}) = \sum_{j=0}^{n} \begin{bmatrix} n \\ j \end{bmatrix} q^{j(j-1)/2} x^j .$$

In Section 2, we demonstrated how to construct a representation of G_n for each m with $0 \leq m \leq n$, by considering the vector space over K whose basis elements are the m-dimensional subspaces of V (the n-dimensional space over $\mathbb{F}_q$ on which G_n acts in the natural way). A way of generalising this idea uses the concept of a <u>composition</u> of n, by which we mean a sequence

$$\lambda = (\lambda_1, \lambda_2, \dots)$$

of non-negative integers, whose sum is n.

Consider subspaces $V_0, V_1, \dots$ of V with the properties that

$$V = V_0 \supseteq V_1 \supseteq V_2 \supseteq \dots \supseteq 0 \text{ , and}$$

$$\dim(V_{i-1}/V_i) = \lambda_i \quad (i \geq 1) .$$

Such a set of subspaces is called a <u>λ-flag</u>. Clearly, the action of G_n on V induces a permutation of the λ-flags, and as a result we may construct a permutation representation of G_n by taking the vector space over K whose basis elements are all the λ-flags. The situation discussed in Section 2 was that where $\lambda = (n - m, m, 0, 0, \dots)$.

It is important to point out that it is unclear whether or not reordering the parts of λ gives rise to an isomorphic permutation representation of G_n. (The result is easy enough if K has characteristic zero, but is not true for arbitrary fields K; see Section 14.) For this reason, it is necessary to work for some time with compositions of n, instead of the more familiar concept of partitions of n. (A <u>partition</u> of n is a composition λ for which $\lambda_1 \geq \lambda_2 \geq \dots$.) This Chapter is devoted to adapting various ideas concerning partitions to suit our needs.

The last non-zero part of the composition λ will always be denoted by λ_h (h standing for "height"), and we shall write

$$(\lambda_1, \lambda_2, \dots, \lambda_h, 0, 0, \dots) = (\lambda_1, \lambda_2, \dots, \lambda_h) .$$

Given a composition λ, its diagram $[\lambda]$ is defined to be the set of ordered pairs (i, j) with $1 \le i \le h$ and $1 \le j \le \lambda_i$, and we always regard $[\lambda]$ as an array of "nodes" as in the example:

$$\lambda = (4, 2, 3) \qquad [\lambda] = \begin{matrix} X & X & X & X \\ X & X & & \\ X & X & X & \end{matrix} .$$

Let λ'_j equal the number of nodes in column j of $[\lambda]$. Then $\lambda' = (\lambda'_1, \lambda'_2, \ldots)$ is a partition of n. For instance, if $\lambda = (4, 2, 3)$ then $\lambda' = (3, 3, 2, 1)$.

4.1 DEFINITIONS. If λ and μ are compositions of n, we write $\lambda \trianglerighteq \mu$ if

$$\sum_{i=1}^{j} \lambda'_i \le \sum_{i=1}^{j} \mu'_i \quad \text{for all } j.$$

We write $\lambda \triangleright \mu$ if $\lambda \trianglerighteq \mu$ but $\mu \ntrianglerighteq \lambda$.

Notice that the relations $\trianglerighteq$ and $\triangleright$ are transitive, but that it is possible to have $\lambda \trianglerighteq \mu$ and $\mu \trianglerighteq \lambda$ without $\lambda = \mu$. In fact, $\lambda \trianglerighteq \mu$ and $\mu \trianglerighteq \lambda$ if and only if μ is obtained from λ by rearranging the parts. For example, if $\lambda = (4, 2, 3)$, $\mu = (3, 2, 4)$, $\nu = (3, 3, 3)$, then $\lambda \trianglerighteq \mu$, $\mu \trianglerighteq \lambda \triangleright \nu$.

A λ-tableau is an array of integers obtained by replacing the nodes in $[\lambda]$ by the numbers $1, 2, \ldots, n$ in some order (no repetitions being allowed). For example,

$$T_1 = \begin{matrix} 1 & 2 & 4 & 5 \\ 3 & 6 & & \\ 7 & 8 & 9 & \end{matrix} \qquad T_2 = \begin{matrix} 3 & 6 & 8 & 5 \\ 2 & 7 & & \\ 1 & 4 & 9 & \end{matrix}$$

are two $(4, 2, 3)$-tableaux. The symmetric group $\mathfrak{S}_n$ acts on the set of λ-tableaux in the natural way (e.g. $T_1(1\,3\,2\,6\,7)(4\,8) = T_2$, above).

If T is a λ-tableau, then row i of T is the set of numbers replacing the nodes (i, j) in $[\lambda]$ $(1 \le j \le \lambda_i)$, and column j of T is the set of numbers replacing the nodes (i, j) in $[\lambda]$ $(1 \le i \le \lambda'_j)$; e.g. $\{4, 9\}$ is column 3 of T_1 above.

Given a tableau T, we define the functions

$$\text{row}_T, \text{col}_T : \{1, 2, \ldots, n\} \to \{1, 2, \ldots\}$$

by

$\text{row}_T(b) = i$ if b belongs to row i of T,

$\text{col}_T(b) = j$ if b belongs to column j of T.

4.2 DEFINITION. The <u>initial λ-tableau</u> will always be denoted by T_λ and it is defined to be the λ-tableau in which the numbers 1, 2, ..., n appear in order when reading from left to right in successive rows.

e.g. If $\lambda = (4, 2, 3)$, then $T_\lambda = \begin{matrix} 1 & 2 & 3 & 4 \\ 5 & 6 & & \\ 7 & 8 & 9 & \end{matrix}$.

5 ROOT SUBGROUPS OF G_n

It is useful to adopt some standard terminology from the theory of root systems. (See, for example, Carter [C].)

Let $\Phi = \{(i, j) \mid i \neq j, 1 \leq i \leq n, 1 \leq j \leq n\}$, and let the symmetric group act on Φ according to the rule:

$$(i, j)\pi = (i\pi, j\pi) \qquad ((i, j) \in \Phi, \pi \in \mathfrak{S}_n) .$$

If $(i, j) \in \Phi$ and $\alpha \in \mathbb{F}_q$, let $x_{ij}(\alpha) \in G_n$ be the matrix with (i, j) entry equal to α, all diagonal entries equal to 1, and zeros elsewhere:

$$x_{ij}(\alpha) = i\begin{pmatrix} 1 & & & \\ & \ddots & & \\ & \alpha & \ddots & \\ & & & 1 \end{pmatrix} \text{ (column } j\text{)} .$$

Then

$$x_{ij}(\alpha_1)x_{ij}(\alpha_2) = x_{ij}(\alpha_1 + \alpha_2) \qquad (\alpha_1, \alpha_2 \in \mathbb{F}_q).$$

Therefore,

$$X_{ij} = \{x_{ij}(\alpha) \mid \alpha \in \mathbb{F}_q\}$$

is a group, isomorphic to $(\mathbb{F}_q, +)$, the additive group of $\mathbb{F}_q$. (X_{ij} is known as a <u>root subgroup</u> of G_n.)

5.1 DEFINITIONS. A subset Γ of Φ is said to be <u>closed</u> if, for all (i, j), $(j, k) \in \Gamma$, we have $(i, k) \in \Gamma$.

If Γ is a closed subset of Φ, let $G(\Gamma)$ be the following subgroup of G_n:

$$G(\Gamma) = \langle X_{ij} \mid (i, j) \in \Gamma\rangle .$$

The next result is very easy, but it is of fundamental importance in the later Chapters:

5.2 THEOREM. If Γ is a closed subset of Φ, then

$$G(\Gamma) = \prod_{(i,j)\in\Gamma} X_{ij} ,$$

where the product can be taken in any order. The group $G(\Gamma)$ consists of all those matrices which have 1's on the diagonal and zeros in those places (i, j) for which $(i, j) \in \Phi \setminus \Gamma$. Once the order of the product has been chosen, each element of $G(\Gamma)$ has a unique expression of the form

$$\prod_{(i,j)\in\Gamma} x_{ij}(\alpha_{ij}) \qquad (\alpha_{ij} \in \mathbb{F}_q) .$$

Proof: This is immediate from the following commutation formulae, which can be deduced from the definition of $x_{ij}(\alpha)$:

(5.3) Suppose that $(i, j), (j, k), (k, \ell) \in \Phi$.

(i) If $(i, \ell) \in \Phi$, then $x_{ij}(\alpha_1)x_{k\ell}(\alpha_2) = x_{k\ell}(\alpha_2)x_{ij}(\alpha_1)$

(ii) If $(i, k) \in \Phi$, then $x_{ij}(\alpha_1)x_{jk}(\alpha_2) = x_{jk}(\alpha_2)x_{ij}(\alpha_1)x_{ik}(\alpha_1\alpha_2)$

$$= x_{ik}(\alpha_1\alpha_2)x_{jk}(\alpha_2)x_{ij}(\alpha_1) .$$

■

In particular, $|G(\Gamma)| = q^{|\Gamma|}$. We also have:

(5.4) If $\Gamma, \Gamma_1, \Gamma_2$ are closed subsets of Φ, with $\Gamma = \Gamma_1 \cup \Gamma_2$ and $\Gamma_1 \cap \Gamma_2$ empty, then $G(\Gamma) = G(\Gamma_1)G(\Gamma_2)$ and every element g of $G(\Gamma)$ can be expressed in a unique way as $g = g_1g_2$ with

$$g_1 \in G(\Gamma_1),\ g_2 \in G(\Gamma_2) .$$

If Γ is a closed subset of Φ, define

$$\Gamma' = \{(i, k) \mid (i, j) \text{ and } (j, k) \in \Gamma \text{ for some } j\} .$$

It is easy to see that Γ' is a closed subset of Φ contained in Γ; indeed, Γ' is strictly contained in Γ unless Γ is empty. From (5.3) we obtain:

(5.5) The commutator subgroup of $G(\Gamma)$ is $G(\Gamma')$.

5.6 DEFINITIONS

(i) $\Phi^+ = \{(i, j) \mid 1 \le i < j \le n\}$ and $U^+ = G(\Phi^+)$

(ii) $\Phi^- = \{(i, j) \mid 1 \le j < i \le n\}$ and $U^- = G(\Phi^-)$.

Then U^+ (respectively, U^-) is the group of upper (respectively, lower) unitriangular matrices:

$$U^+ = \text{the set of matrices like } \begin{pmatrix} 1 & & * \\ & \ddots & \\ 0 & & 1 \end{pmatrix},$$

$$U^- = \text{the set of matrices like } \begin{pmatrix} 1 & & 0 \\ & \ddots & \\ * & & 1 \end{pmatrix}.$$

The commutator subgroup $(U^-)'$ of U^- consists of those lower unitriangular matrices which have zero immediately below the diagonal:

$$(U^-)' = \text{the set of matrices like } \begin{pmatrix} 1 & & & & & 0 \\ 0 & 1 & & & & \\ & 0 & 1 & & & \\ & & & \ddots & & \\ & & & & \ddots & \\ * & & & & 0 & 1 \end{pmatrix}$$

If $1 \le i \le n$ and β is a non-zero element of $\mathbb{F}_q$, let $h_i(\beta)$ be the matrix in G_n with (i, i) entry equal to β, all other diagonal entries equal to 1 and zeros elsewhere:

$$h_i(\beta) = \begin{array}{c} \\ \\ \\ \\ i \\ \\ \\ \\ \end{array} \overset{\displaystyle i}{\begin{pmatrix} 1 & & & & & & & \\ & 1 & & & & & 0 & \\ & & \ddots & & & & & \\ & & & \beta & & & & \\ & & & & \ddots & & & \\ & 0 & & & & \ddots & & \\ & & & & & & & 1 \end{pmatrix}}$$

Let $H = \langle h_i(\beta) \mid 1 \le i \le n,\ \beta \in \mathbb{F}_q \setminus \{0\}\rangle$. Thus, H is the group of diagonal matrices:

$$H = \text{the set of matrices like } \begin{pmatrix} * & & & 0 \\ & * & & \\ & & \ddots & \\ 0 & & & * \end{pmatrix}.$$

For $\alpha, \beta \in \mathbb{F}_q$, $\beta \neq 0$, we have

$$(5.7) \quad \begin{aligned} h_k^{-1}(\beta) x_{ij}(\alpha) h_k(\beta) &= x_{ij}(\alpha\beta), \quad \text{if } k = j, \\ &= x_{ij}(\alpha\beta^{-1}), \quad \text{if } k = i, \\ &= x_{ij}(\alpha), \quad \text{if } k \neq i, j. \end{aligned}$$

Thus, H normalizes each root subgroup. Hence H normalizes U^+ and U^-. The groups $U^+H = HU^+$ and $U^-H = HU^-$ are denoted by B^+ and B^-, respectively, and are called <u>Borel subgroups</u>.

$$B^+ = \text{the set of matrices like } \begin{pmatrix} * & & & * \\ & * & & \\ & & \ddots & \\ 0 & & & * \end{pmatrix},$$

$$B^- = \text{the set of matrices like } \begin{pmatrix} * & & & 0 \\ & * & & \\ & & \ddots & \\ * & & & * \end{pmatrix}.$$

By (5.7), if $1 \leq d < n$, then

$$h_d^{-1}(\beta) x_{d+1,d}(\alpha) h_d(\beta) = x_{d+1,d}(\alpha\beta)$$

and

$$h_d^{-1}(\beta) x_{j+1,j}(\alpha) h_d(\beta) = x_{j+1,j}(\alpha) \quad \text{if } j > d.$$

Hence:

(5.8) Suppose that $1 \leq d < n$, and that β is a non-zero element of $\mathbb{F}_q$. Then there exists $h \in H$, such that for all $\alpha \in \mathbb{F}_q$,

$$h^{-1} x_{d+1,d}(\alpha) h = x_{d+1,d}(\alpha\beta), \quad \text{and}$$

$$h^{-1} x_{j+1,j}(\alpha) h = x_{j+1,j}(\alpha) \quad \text{for all } j \neq d.$$

If $\pi \in \mathfrak{S}_n$, we shall identify π with the permutation matrix in G_n which acts on the basis $e_1, e_2, \ldots, e_n$ of V by

$$e_1\pi = e_{i\pi} \qquad (1 \leq i \leq n) .$$

Let W be the subgroup of G_n obtained by this identification:

$$W = \text{the set of permutation matrices.}$$

For calculations, it is useful to notice that if $g \in G_n$ and $\pi \in W$, then $\pi^{-1}g$ (respectively, $g\pi$) is obtained from g by applying the permutation π to the rows (respectively, columns) of g. For example,

$$(5.9) \quad \begin{pmatrix} 0 & 1 & 0 & 0 \\ 0 & 0 & 1 & 0 \\ 1 & 0 & 0 & 0 \\ 0 & 0 & 0 & 1 \end{pmatrix} \begin{pmatrix} 1 & 0 & 0 & 0 \\ \alpha & 1 & 0 & 0 \\ \beta & \gamma & 1 & 0 \\ \delta & \varepsilon & \zeta & 1 \end{pmatrix} = \begin{pmatrix} \alpha & 1 & 0 & 0 \\ \beta & \gamma & 1 & 0 \\ 1 & 0 & 0 & 0 \\ \delta & \varepsilon & \zeta & 1 \end{pmatrix} .$$

The following results are elementary:

(5.10) (i) For each $\pi \in W$, $\alpha \in \mathbb{F}_q$, $(i, j) \in \Phi$,

$$\pi^{-1}x_{ij}(\alpha)\pi = x_{i\pi,j\pi}(\alpha) \text{ , and}$$

$$\pi^{-1}X_{ij}\pi = X_{i\pi,j\pi} .$$

(ii) W normalizes H.

6 SUBGROUPS OF G_n ASSOCIATED WITH COMPOSITIONS

We shall study the KG_n-module which consists of the vector space over K whose basis elements are the λ-flags. It is often best, in practice, to view this module as a right ideal of the group algebra KG_n. To do this, we consider the stabilizer of the λ-flag

$$V = V_0 \supseteq V_1 \supseteq V_2 \supseteq \cdots \supseteq V_h = 0 ,$$

where

$$V_i = \langle e_{\Lambda_i+1}, e_{\Lambda_i+2}, \ldots , e_n \rangle \qquad (0 \le i \le h) ,$$

and

$$\Lambda_i = \lambda_1 + \lambda_2 + \ldots + \lambda_i .$$

The stabilizer of this λ-flag is

$$P_\lambda = \text{the set of matrices like } \begin{pmatrix} * & & * \\ & * & \\ 0 & & \ddots \end{pmatrix} .$$

(with diagonal blocks of sizes $\lambda_1, \lambda_2, \ldots$)

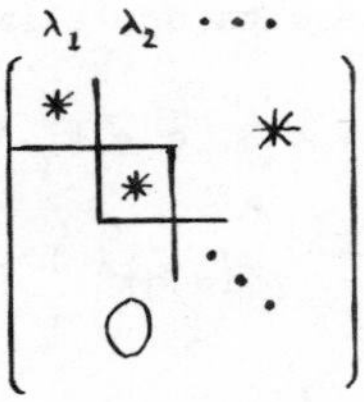

Recalling the definition of the initial λ-tableau T_λ, we may write P_λ (a so-called <u>parabolic subgroup</u> of G_n) as

$$P_\lambda = \langle H, X_{ij} \mid \text{row}_{T_\lambda}(i) \le \text{row}_{T_\lambda}(j) \rangle .$$

We shall be amply rewarded later for using this apparently clumsy notation involving T_λ.

6.1 DEFINITIONS

(i) Let $U_\lambda^+ = \langle X_{ij} \mid \text{row}_{T_\lambda}(i) < \text{row}_{T_\lambda}(j) \rangle$

(ii) Let $U_\lambda^- = \langle X_{ij} \mid \text{row}_{T_\lambda}(i) > \text{row}_{T_\lambda}(j) \rangle$.

Then

$$U_\lambda^+ = \text{the set of matrices like } \begin{array}{c} \lambda_1 \;\; \lambda_2 \; \cdots \qquad \\ \left[\begin{array}{ccc} I & & * \\ & I & \\ 0 & & \ddots \end{array}\right] \end{array}.$$

$$U_\lambda^- = \text{the set of matrices like } \begin{array}{c} \lambda_1 \;\; \lambda_2 \; \cdots \qquad \\ \left[\begin{array}{ccc} I & & 0 \\ & I & \\ * & & \ddots \end{array}\right] \end{array}.$$

For example, when $\lambda = (1, 1, 1, \ldots)$, $P_\lambda = B^+$, $U_\lambda^+ = U^+$, $U_\lambda^- = U^-$.

If Γ is any closed subset of Φ and $\pi \in \mathfrak{S}_n$, then

$$\Gamma_1 = \{(i, j) \in \Gamma \mid \operatorname{row}_{T_\lambda \pi}(i) \le \operatorname{row}_{T_\lambda \pi}(j)\}$$

$$\Gamma_2 = \{(i, j) \in \Gamma \mid \operatorname{row}_{T_\lambda \pi}(i) > \operatorname{row}_{T_\lambda \pi}(j)\}$$

are disjoint closed subsets of Φ whose union is Γ. Therefore, applying (5.4) to the case where $\Gamma = \Phi^-$, we have

(6.2) For all $\pi \in \mathfrak{S}_n$, $U^- = G(\Gamma_1)G(\Gamma_2)$, where

$$\Gamma_1 = \{(i, j) \mid i > j \text{ and } \operatorname{row}_{T_\lambda \pi}(i) \le \operatorname{row}_{T_\lambda \pi}(j)\}$$

$$\Gamma_2 = \{(i, j) \mid i > j \text{ and } \operatorname{row}_{T_\lambda \pi}(i) > \operatorname{row}_{T_\lambda \pi}(j)\}$$

and $G(\Gamma_1) \cap G(\Gamma_2) = 1$. That is

$$U^- = (U^- \cap \pi^{-1} P_\lambda \pi)(U^- \cap \pi^{-1} U_\lambda^- \pi) .$$

Since we shall consider the permutation representation of G_n on the right cosets of P_λ, we wish to determine a set of right coset representatives for P_λ in G_n. (By a right coset, we mean a set of the form $P_\lambda g$ with $g \in G_n$.) We shall illustrate with several examples of these coset representatives, since they will be used frequently in later chapters.

First, we deal with the special case where $\lambda = (1, 1, 1, \ldots)$:

7.1 THEOREM. Every right coset of B^+ in G_n has the form $B^+\pi u$, where $\pi \in W$ and $u \in U^- \cap \pi^{-1}U^-\pi$.

Proof: It is known that G_n is a union of double cosets $B^+\pi B^+$, where $\pi \in W$. But B^+ and B^- are conjugate by the element of W which reverses the numbers $1, 2, \ldots, n$. Therefore,

$$G_n = \bigcup_{\pi \in W} B^+\pi B^-$$

(we shall call this "the Bruhat Decomposition"). Now, every element of B^- has the form hu with $h \in H$, $u \in U^-$, and W normalizes $H \subseteq B^+$. Thus, every right coset of B^+ in G_n has the form $B^+\pi u$, with $u \in U^-$. Further, by (6.2), we may write $u = u'u''$, where

$$u' \in U^- \cap \pi^{-1}B^+\pi , \qquad u'' \in U^- \cap \pi^{-1}U^-\pi .$$

Then

$$B^+\pi u = B^+\pi u'\pi^{-1}\pi u'' = B^+\pi u'' ,$$

as stated. ∎

We remark that it is simple enough to construct an alternative proof, and at the same time prove the Bruhat Decomposition. For, given a right coset B^+g of B^+ in G_n, then we can pre-multiply g by elements of B^+ in such a way as to "row reduce g, starting at the bottom" (cf. (2.11) and the argument following it). We leave the details to the reader.

7.2 EXAMPLE. If $n = 3$, then each right coset of B^+ in G_n has one of the following forms:

$$\begin{pmatrix} 1 & 0 & 0 \\ \alpha & 1 & 0 \\ \beta & \gamma & 1 \end{pmatrix}, \begin{pmatrix} 0 & 1 & 0 \\ 1 & 0 & 0 \\ \alpha & \beta & 1 \end{pmatrix}, \begin{pmatrix} 1 & 0 & 0 \\ \beta & 0 & 1 \\ \alpha & 1 & 0 \end{pmatrix}, \begin{pmatrix} 0 & 1 & 0 \\ 0 & \alpha & 1 \\ 1 & 0 & 0 \end{pmatrix}, \begin{pmatrix} 0 & 0 & 1 \\ 1 & 0 & 0 \\ \alpha & 1 & 0 \end{pmatrix}, \begin{pmatrix} 0 & 0 & 1 \\ 0 & 1 & 0 \\ 1 & 0 & 0 \end{pmatrix}.$$

$(\alpha, \beta, \gamma \in \mathbb{F}_q)$.

The element $\begin{pmatrix} \beta & 0 & 1 \\ \alpha & 1 & 0 \\ 1 & 0 & 0 \end{pmatrix}$ is in the right coset $B^+ \begin{pmatrix} 0 & 0 & 1 \\ 0 & 1 & 0 \\ 1 & 0 & 0 \end{pmatrix}$,

since we can "row reduce it starting from the bottom". Explicitly,

$$\begin{pmatrix} 1 & 0 & -\beta \\ 0 & 1 & -\alpha \\ 0 & 0 & 1 \end{pmatrix} \begin{pmatrix} \beta & 0 & 1 \\ \alpha & 1 & 0 \\ 1 & 0 & 0 \end{pmatrix} = \begin{pmatrix} 0 & 0 & 1 \\ 0 & 1 & 0 \\ 1 & 0 & 0 \end{pmatrix}.$$

It is interesting to verify that $q^3 + 2q^2 + 2q + 1 = (1 + q)(1 + q + q^2)$ $= |G_n : B^+|$. ∎

We should like to comment here on our notation. In Theorem 7.1 we have committed ourselves to working with right cosets; there was also a choice of signs ± in the double coset decomposition

$$G_n = \bigcup_{\pi \in W} B^{\pm} \pi B^{\pm}.$$

The alternatives involve thinking about column vectors instead of row vectors, and reversing the parts in partitions. All our later results could be amended to cope with whatever form of Theorem 7.1 we cared to take, but we feel that the development runs most smoothly with the choice we have made.

Consider now the right cosets of the parabolic subgroup P_λ of G_n. Since $B^+ \subseteq P_\lambda$, each right coset representative of P_λ has the form πu with $\pi \in W$, $u \in U^-$.

7.3 LEMMA. *If $P_\lambda \pi_1 u_1 = P_\lambda \pi_2 u_2$, with $\pi_1, \pi_2 \in W$, and $u_1, u_2 \in U^-$, then $\pi_1 \pi_2^{-1} \in W \cap P_\lambda$.*

Proof: If $P_\lambda \pi_1 u_1 = P_\lambda \pi_2 u_2$, then $\pi_1 u_1 u_2^{-1} \pi_2^{-1} \in P_\lambda$. But for all i, the matrix

$u_1u_2^{-1}$ has a 1 in the (i, i) place, and so $\pi_1u_1u_2^{-1}\pi_2^{-1}$ has a 1 in each place where $\pi_1\pi_2^{-1}$ has a 1. Therefore, $\pi_1\pi_2^{-1} \in W \cap P_\lambda$. ∎

The Lemma illustrates nicely the analogy between our parabolic subgroups of G_n and Young subgroups of $\mathfrak{S}_n$.

We know from (6.2) that, given $\pi \in W$, every $u \in U^-$ can be written in the form

$$u = u_1u_2, \text{ with } u_1 \in U^- \cap \pi^{-1}P_\lambda\pi \text{ and } u_2 \in U^- \cap \pi^{-1}U_\lambda^-\pi.$$

Therefore,

(7.4) If $\pi \in W$, $u \in U^-$, then $P_\lambda\pi u = P_\lambda\pi u_2$ for some $u_2 \in U^- \cap \pi^{-1}U_\lambda^-\pi$.

Remembering that $(\pi^{-1}P_\lambda\pi) \cap (\pi^{-1}U_\lambda^-\pi) = 1$, we have proved:

7.5 THEOREM <u>Suppose that λ is a composition of n. Then every right coset of P_λ in G_n has the form</u>

$$P_\lambda\pi u,$$

<u>where $\pi \in W$, and $u \in U^- \cap \pi^{-1}U_\lambda^-\pi = G(\Gamma)$, and</u>

$$\Gamma = \{(i, j) \mid i > j \text{ and } \mathrm{row}_{T_\lambda\pi}(i) > \mathrm{row}_{T_\lambda\pi}(j)\}.$$

<u>Furthermore, if</u>

$$P_\lambda\pi_1u_1 = P_\lambda\pi_2u_2,$$

<u>with $\pi_1, \pi_2 \in W$, and $u_1, u_2 \in U^- \cap \pi^{-1}U_\lambda^-\pi$, then</u>

$$\pi_1\pi_2^{-1} \in W \cap P_\lambda, \text{ and } u_1 = u_2.$$ ∎

It is useful to understand clearly the form of the right coset representatives πu of P_λ. Suppose π is given; this is a permutation matrix, so each row and each column contains precisely one 1. The elements πu, with $u \in U^-$, are obtained by replacing each 0 which lies to the left of a 1 by an arbitrary element of $\mathbb{F}_q$ (cf. (5.9)). In order to explain which of these elements πu have $u \in U^- \cap \pi^{-1}U_\lambda^-\pi$, we imagine that the rows of the matrix π are divided into batches, batch b consisting of rows

$$1 + \sum_{i=1}^{b-1} \lambda_i \quad \text{to} \quad \sum_{i=1}^{b} \lambda_i .$$

Then the elements in $\pi\,(U^- \cap \pi^{-1}U_\lambda^-\pi)$ are obtained by replacing by an arbitrary element of $\mathbb{F}_q$ each 0 in π which lies both to the left of a 1 in the same row and below a 1 which lies in a stictly higher batch of rows. (cf. Example 7.2; there each individual row forms a batch.)

7.6 EXAMPLES.

(i) If $n = 8$, $\lambda = (2, 3, 3)$ and

$$\pi = \begin{pmatrix} &&&&1&&& \\ 1&&&&&&& \\ \hline &&1&&&&& \\ &1&&&&&& \\ &&&&&&1& \\ \hline &&&&&1&& \\ &&&1&&&& \\ &&&&&&&1 \end{pmatrix}$$

then every element of $\pi\,(U^- \cap \pi^{-1}U_\lambda^-\pi)$ has the form

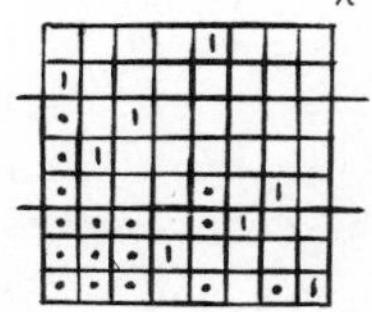

where each . is replaced by an element of $\mathbb{F}_q$. (Empty places denote zero.)

In this example,

$$T_\lambda = \begin{matrix} 1 & 2 & \\ 3 & 4 & 5 \\ 6 & 7 & 8 \end{matrix} \qquad T_\lambda\pi = \begin{matrix} 5 & 1 & \\ 3 & 2 & 7 \\ 6 & 4 & 8 \end{matrix}$$

and $U^- \cap \pi^{-1}U_\lambda^-\pi$ is the set of matrices of the form

$$\begin{pmatrix} 1&&&&&&& \\ \cdot&1&&&&&& \\ \cdot&&1&&&&& \\ \cdot&\cdot&\cdot&1&&&& \\ &&&&1&&& \\ \cdot&\cdot&\cdot&&\cdot&1&& \\ \cdot&&&&\cdot&&1& \\ \cdot&\cdot&\cdot&&\cdot&&\cdot&1 \end{pmatrix} .$$

In this last matrix, there is a . in the (i, j) place if $i > j$ and $row_{T_\lambda \pi}(i) > row_{T_\lambda \pi}(j)$.

(ii) If $n = 4$ and $\lambda = (2, 2)$, then the right coset representatives of P_λ in G_n are

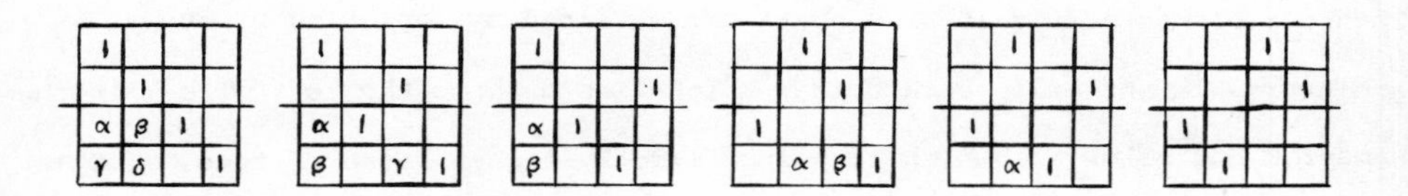

$(\alpha, \beta, \gamma, \delta \in \mathbb{F}_q)$.

Whenever $\lambda = (\lambda_1, \lambda_2)$ has two non-zero parts, a coset representative is determined by the entries in the last λ_2 rows (compare Example 7.6(ii) with Example 2.13). We recall that each right coset of P_λ in G_n always corresponds to a λ-flag; all that Theorem 7.5 achieves, therefore, is a "coding" for the collection of λ-flags. For example, if $\lambda = (2, 3, 3)$, the right coset representatives of P_λ in G_n correspond to flags

$$V = V_0 \supset V_1 \supset V_2 \supset V_3 = 0$$

where $\dim V_0 = 8$, $\dim V_1 = 6$, $\dim V_3 = 3$.

Notice that the last λ_h rows of the coset representative matrix correspond to a subspace V_{h-1} of V, the last $\lambda_{h-1} + \lambda_h$ rows to a subspace $V_{h-2} \supseteq V_{h-1}$, and so forth. The fact that the batch of λ_{h-1} rows correspond to a quotient space V_{h-2}/V_{h-1}, and so on, illustrates our point about "row reducing from the bottom".

To summarize, there is nothing mysterious about the coset representat constructed in Theorem 7.5; they merely represent a translation of the concep of λ-flags into a language which allows us to work with the group algebra of G_n.

8 SUBGROUPS OF G_n USED FOR INDUCTION

Unlike the situation for the symmetric groups, where $\mathfrak{S}_{n-1}$ is clearly the subgroup of $\mathfrak{S}_n$ which should be used for induction on n, there is some choice of "obvious" subgroups of G_n. We can do better than restricting our attention to a subgroup isomorphic to $GL_{n-1}(q)$, and we introduce here the subgroups we shall use. In the next section we shall derive, using them, some results on induced representations. (Hence the ambiguous title of this chapter!)

There are four types of subgroup which we wish to name G_r, G_r^*, H_r^*, A_r, respectively, and these are most clearly described in terms of pictures:

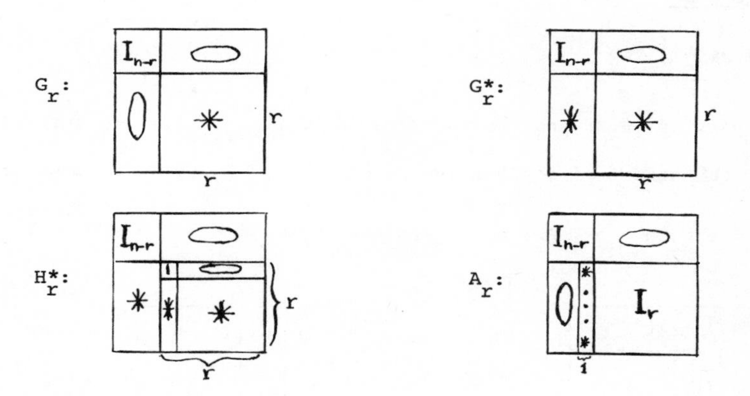

To avoid any possible ambiguity, we give formal definitions of these groups.

8.1 DEFINITION. Assume $0 \le r \le n$.

(i) $G_r = \langle X_{ij}, h_i(\beta) \mid i, j > n - r, \beta \in \mathbb{F}_q \setminus \{0\}\rangle$

(ii) $G_r^* = \langle X_{ij}, h_i(\beta) \mid i > n - r, \beta \in \mathbb{F}_q \setminus \{0\}\rangle$

(iii) $H_r^* = \langle X_{ij}, h_k(\beta) \mid i, k > n - r + 1, \beta \in \mathbb{F}_q \setminus \{0\},$ or $i = n - r + 1$ and $j \le n - r\rangle$

(iv) $A_r = \langle X_{i,n-r} \mid i > n - r\rangle$.

In particular, $G_0 = G_0^* = H_0^* = A_0$ = the identity subgroup.

We have (see Definition 6.1 and (5.3)):

(8.2) $|A_r| = q^r$.

$G_{r-1}^* \lhd H_r^* \subset G_r^*$.

$|H_r^* : G_{r-1}^*| = q^{n-r}$ and $|G_r^* : H_r^*| = q^r - 1$.

$U^-_{(n-r,r)}$ is an abelian normal subgroup of G_r^*.

$U^-_{(n-r,r)} \cap G_r = 1$ and $G_r^* = U^-_{(n-r,r)} G_r$.

8.3 THEOREM *$A_r \lhd G_r^*$, and given any two non-identity elements $a_1, a_2 \in A_r$, there exists $g \in G_r$ such that*

$a_2 = g a_1 g^{-1}$.

Proof: In view of (8.2) and the fact that $A_r \subseteq U^-_{(n-r,r)}$, it is sufficient to consider the effect of conjugating an element of A_r by an element of G_r.

Write elements $a \in A_r$, $g \in G_r$ in the form

$$a = \left(\begin{array}{c|c} I_{n-r} & 0 \\ \hline 0 \;\vert\; \underset{\sim}{u} & I_r \end{array}\right) \qquad g = \left(\begin{array}{c|c} I_{n-r} & 0 \\ \hline 0 & m \end{array}\right)$$

where $\underset{\sim}{u}$ is a column vector of height r and m is a non-singular $r \times r$ matrix. Then direct calculation reveals that

$$gag^{-1} = \left(\begin{array}{c|c} I_{n-r} & 0 \\ \hline 0 \;\vert\; \underset{\sim}{v} & I_r \end{array}\right)$$

where $\underset{\sim}{v}$ is the column vector $m\underset{\sim}{u}$. The result now follows. ∎

From now on (except where otherwise explicitly stated in one or two examples), we assume that K is a field whose characteristic does not divide q.

If q is a power of the prime p, let $\bar{K}$ be the field obtained from K by adjoining a primitive p^{th} root of unity.

8.4 DEFINITION. If A is a subgroup of G_n, then a linear $\bar{K}$-character χ of A is a mapping

$$\chi: \quad A \to \bar{K} \setminus \{0\}$$

such that $\chi(a_1 \cdot a_2) = \chi(a_1)\chi(a_2)$ for all $a_1, a_2 \in A$. A linear $\bar{K}$-character χ of A is said to be non-trivial if $\chi(a) \neq 1$ for some $a \in A$.

Now, every element a of A_r is determined by the column vector $\underset{\sim}{u}$ of height r in the description

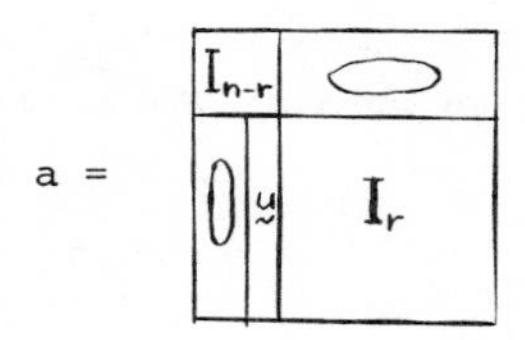

and the effect of multiplying two elements of A_r is to add the corresponding column vectors. Hence the proof of Theorem 8.3 shows that the action of G_r on A_r, by conjugation, is the same as the action of G_r (viewed as the group of $r \times r$ non-singular matrices over $\mathbb{F}_q$) on the set V_r of column vectors with height r and entries from $\mathbb{F}_q$. If $q = p^t$, then V_r is a vector space (of dimension tr) over the field $\mathbb{F}_p$ of p elements. Since G_r acts transitively on the set of non-zero elements of the $\mathbb{F}_p$-vector space V_r, G_r also acts transitively on the set of non-zero elements of the $\mathbb{F}_p$-vector space V_r^* dual to V_r. (As usual, V_r^* is the set of $\mathbb{F}_p$-linear mappings θ from V_r to $\mathbb{F}_p$, and the G_r action on this set is defined by $v(\theta g) = g(v\theta)$ for $v \in V_r$, $g \in G_r$.) This proves:

8.5 THEOREM. *If ϕ_1, ϕ_2 are any two non-trivial linear $\bar{K}$-characters of A_r then for some $g \in G_r$,*

$$\phi_1(gag^{-1}) = \phi_2(a) \quad \text{for all } a \in A_r.$$

Since $(\mathbb{F}_q, +)$ is an elementary abelian p-group, it has q distinct linear $\bar{K}$-characters, by which we mean mappings

$$\chi: \mathbb{F}_q \to \bar{K} \setminus \{0\}$$

such that $\chi(\alpha_1 \cdot \alpha_2) = \chi(\alpha_1)\chi(\alpha_2)$ for all $\alpha_1, \alpha_2 \in \mathbb{F}_q$. A particular case of Theorem 8.5 (arising from taking r = 1) is:

8.6 COROLLARY. *If χ_1, χ_2 are any two non-trivial linear $\bar{K}$-characters of $(\mathbb{F}_q, +)$, then for some $\beta \in \mathbb{F}_q \setminus \{0\}$,*

$$\chi_1(\alpha\beta) = \chi_2(\alpha) \quad \text{for all } \alpha \in \mathbb{F}_q.$$

Given any linear $\bar{K}$-character χ of $(\mathbb{F}_q, +)$, we obtain from it a linear $\bar{K}$-character ϕ of A_r, by defining

$$\phi: \left(\begin{array}{c|cc} I_{n-r} & & 0 \\ \hline & \alpha_1 & \\ 0 & \vdots & I_r \\ & \alpha_r & \end{array}\right) \mapsto \chi(\alpha_1).$$

8.7 LEMMA. *If the character ϕ of A_r, described above, is non-trivial, th*

$$H_r^* = \{g \in G_r^* \mid \phi(gag^{-1}) = \phi(a) \quad \text{for all } a \in A_r\} .$$

Proof: Since

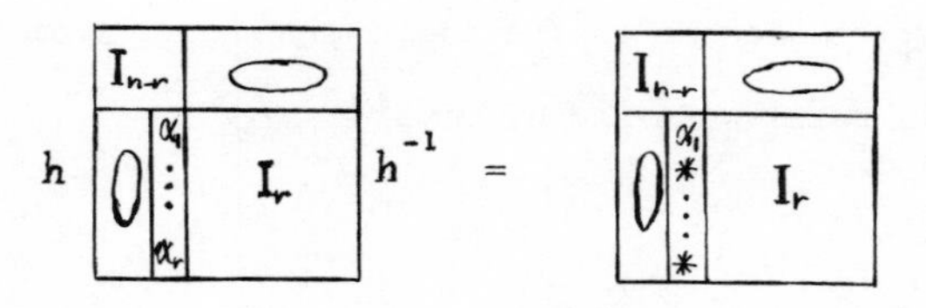

for all $h \in H^*_r$, we certainly have

$$H^*_r \subseteq \{g \in G^*_r \mid \phi(gag^{-1}) = \phi(a) \quad \text{for all } a \in A_r\} .$$

But, by Theorem 8.5, G^*_r is transitive on the $q^r - 1$ non-trivial linear $\bar{K}$-characters of A_r. Since $|G^*_r : H^*_r| = q^r - 1$, we deduce the result. ■

9 SOME IDEMPOTENT ELEMENTS OF $\bar{K}G_n$

With the preliminaries now behind us, we commence our study of $\bar{K}G_n$-modules in this chapter. For each closed subset Γ of Φ we shall define several idempotent elements of $\bar{K}G_n$, and investigate the properties of these idempotents.

9.1 DEFINITIONS. Let $\chi_1, \ldots, \chi_q$ be the distinct linear $\bar{K}$-characters of $(\mathbb{F}_q, +)$, with χ_1 equal to the trivial character (that is, $\chi_1(\alpha) = 1$ for all $\alpha \in \mathbb{F}_q$).

Given a closed subset Γ of Φ, let c be any function from Γ to $\{1, 2, \ldots, q\}$ such that

$$(i, j)c = 1 \quad \text{if } (i, j) \in \Gamma'.$$

Define the function χ_c from $G(\Gamma)$ to $\bar{K}$ by

$$\chi_c\colon \quad \prod_{(i,j)\in\Gamma} x_{ij}(\alpha_{ij}) \mapsto \prod_{(i,j)\in\Gamma} \chi_{(i,j)c}(\alpha_{ij}), \qquad (\alpha_{ij} \in \mathbb{F}_q).$$

In view of Theorem 5.2, and the fact that we have taken c to have value 1 on Γ', χ_c is a well-defined linear $\bar{K}$-character of $G(\Gamma)$. Hence

$$\frac{1}{q^{|\Gamma|}} \sum_{g\in G(\Gamma)} \chi_c(g)g$$

is an idempotent element of $\bar{K}G_n$.

9.2 THEOREM. <u>If Γ is a closed subset of Φ, and $\Gamma_1, \Gamma_2, \ldots, \Gamma_k$ are disjoint closed subsets of Φ whose union is Γ, then</u>

$$\frac{1}{q^{|\Gamma|}} \sum_{g\in G(\Gamma)} \chi_c(g)g = \prod_{i=1}^{k} \left(\frac{1}{q^{|\Gamma_i|}} \sum_{g\in G(\Gamma_i)} \chi_c(g)g \right),$$

<u>where the product can be taken in any order.</u>

<u>Proof:</u> Making repeated use of Theorem 5.2, we have

$$\frac{1}{q^{|\Gamma|}} \sum_{g \in G(\Gamma)} \chi_c(g) g$$

$$= \frac{1}{q^{|\Gamma|}} \sum_f \left\{ \chi_c \left(\prod_{(i,j) \in \Gamma} x_{ij}(f(i, j)) \right) \prod_{(i,j) \in \Gamma} x_{ij}(f(i, j)) \right\},$$

the sum being over all functions f from Γ to $\mathbb{F}_q$, and the product being in any order,

$$= \frac{1}{q^{|\Gamma|}} \sum_f \left\{ \prod_{(i,j) \in \Gamma} \chi_c(x_{ij}(f(i, j))) x_{ij}(f(i, j)) \right\}$$

$$= \frac{1}{q^{|\Gamma|}} \prod_{(i,j) \in \Gamma} \left[\sum_{\alpha \in \mathbb{F}_q} \chi_{(i,j)c}(\alpha)\ x_{ij}(\alpha) \right]$$

$$= \prod_{(i,j) \in \Gamma} \left[\frac{1}{q} \sum_{g \in X_{ij}} \chi_c(g) g \right].$$

Since the product here can be taken in any order we please, the Theorem follows. ■

9.3 EXAMPLE. Let $n = 3$, $q = 2$, $\Gamma = \Phi^-$, and c be defined by

$$c:\ (2, 1) \mapsto 2$$
$$(3, 2) \mapsto 2$$
$$(3, 1) \mapsto 1.$$

Then $\sum_{g \in G(\Gamma)} \chi_c(g) g = \sum_{\alpha,\beta,\gamma \in \mathbb{F}_2} \chi_2(\alpha)\chi_2(\gamma) \begin{pmatrix} 1 & 0 & 0 \\ \alpha & 1 & 0 \\ \beta & \gamma & 1 \end{pmatrix}$

$$= \left\{ \begin{pmatrix} 1 & 0 & 0 \\ 0 & 1 & 0 \\ 0 & 0 & 1 \end{pmatrix} + \begin{pmatrix} 1 & 0 & 0 \\ 0 & 1 & 0 \\ 1 & 0 & 1 \end{pmatrix} \right\} \left\{ \begin{pmatrix} 1 & 0 & 0 \\ 0 & 1 & 0 \\ 0 & 0 & 1 \end{pmatrix} - \begin{pmatrix} 1 & 0 & 0 \\ 1 & 1 & 0 \\ 0 & 0 & 1 \end{pmatrix} \right\} \left\{ \begin{pmatrix} 1 & 0 & 0 \\ 0 & 1 & 0 \\ 0 & 0 & 1 \end{pmatrix} - \begin{pmatrix} 1 & 0 & 0 \\ 0 & 1 & 0 \\ 0 & 1 & 1 \end{pmatrix} \right\}.$$

The result is the same if we take the product in any other order. However, notice that the last two terms in the product do not commute.

9.4 DEFINITIONS. Suppose that $1 \le r \le n$.

(i) Let $\Gamma(r) = \{(i, j) \in \Phi^- \mid j \le r\}$.

(ii) Define the function c_r from $\Gamma(r)$ to $\{1, 2, \ldots, q\}$ by

$$(j + 1, j)c_r = 2 \text{ if } j < r$$

$$(i, j)c_r = 1 \text{ for all other } (i, j) \in \Gamma(r).$$

(iii) Let

$$E_r = \frac{1}{q^{|\Gamma(r)|}} \sum_{g \in G(\Gamma(r))} \chi_{c_r}(g)g \ .$$

9.5 EXAMPLE. If $n = 3$, then

$$E_1 = \frac{1}{q^2} \sum_{\alpha,\beta \in \mathbb{F}_q} \begin{pmatrix} 1 & 0 & 0 \\ \alpha & 1 & 0 \\ \beta & 0 & 1 \end{pmatrix}$$

$$E_2 = \frac{1}{q^3} \sum_{\alpha,\beta,\gamma \in \mathbb{F}_q} \chi_2(\alpha) \begin{pmatrix} 1 & 0 & 0 \\ \alpha & 1 & 0 \\ \beta & \gamma & 1 \end{pmatrix}$$

$$E_3 = \frac{1}{q^3} \sum_{\alpha,\beta,\gamma \in \mathbb{F}_q} \chi_2(\alpha)\chi_2(\gamma) \begin{pmatrix} 1 & 0 & 0 \\ \alpha & 1 & 0 \\ \beta & \gamma & 1 \end{pmatrix} .$$

We emphasize that these are elements of $\bar{K}G_3$. (We never consider matrix sums

From (5.8) and Corollary 8.6 we have the following useful result:

(9.6) Suppose that $1 \le d < n$. Let ϕ_d, ϕ_d' be non-trivial linear $\bar{K}$-character of $X_{d+1,d}$, and for each j with $1 \le j < n$ and $j \ne d$, let ϕ_j be an arbitrary linear $\bar{K}$-character of $X_{j+1,j}$.

Then, for some $h \in H$,

$$h^{-1}\Big(\sum_{g \in X_{d+1,d}} \phi_d(g)g\Big)h = \sum_{g \in X_{d+1,d}} \phi_d'(g)g, \text{ and}$$

$$h^{-1}\Big(\sum_{g \in X_{j+1,j}} \phi_j(g)g\Big)h = \sum_{g \in X_{j+1,j}} \phi_j(g)g, \text{ for all } j \ne d.$$

Applying this, we obtain:

(9.7) Suppose that c_r' is any function from $\Gamma(r)$ to $\{1, 2, \ldots, q\}$ such that

$$(j + 1, j)c_r' > 1 \text{ if } j < r$$

$$(i, j)c_r' = 1 \text{ for all other } (i, j) \in \Gamma(r).$$

Then, for some $h \in H$,

$$h^{-1}E_r h = \frac{1}{q^{|\Gamma(r)|}} \sum_{g \in G(\Gamma(r))} \chi_{c'_r}(g)g.$$

For this reason, the exact nature of c_r does not matter for our purposes - any function c'_r satisfying the conditions of (9.7) would do equally well in the definition of E_r.

Now, $G(\Gamma(r)) = A_{n-1} A_{n-2} \dots A_{n-r}$ (see Definition 8.1). For each s with $0 \le s \le n - 1$, A_s is an elementary abelian p group (q = a power of p), of order q^s.

9.8 DEFINITIONS. Suppose that $0 \le s \le n - 1$.

(i) Let $\phi_1, \phi_2, \dots, \phi_{q^s}$ be the linear $\bar{K}$-characters of A_s, the notation being chosen so that for $1 \le i \le q$, ϕ_i is the character given by

$$\phi_i : \left(\begin{array}{c|c} I_{n-s} & 0 \\ \hline \begin{array}{c|c} 0 & \begin{matrix}\alpha_1 \\ \vdots \\ \alpha_s\end{matrix} \end{array} & I_s \end{array}\right) \mapsto \chi_i(\alpha_1).$$

(ii) For $1 \le i \le q^s$, let

$$A_s^{(i)} = \frac{1}{q^s} \sum_{g \in A_s} \phi_i(g)g.$$

Theorem 9.2 gives

$$E_r = A_{n-1}^{(2)} A_{n-2}^{(2)} \dots A_{n-r+1}^{(2)} A_{n-r}^{(1)}. \tag{9.9}$$

In particular, $E_1 = A_{n-1}^{(1)}$ and $E_n = A_{n-1}^{(2)} \dots A_1^{(2)}$.

Now, $\{A_{n-1}^{(i)} \mid 1 \le i \le q^{n-1}\}$ is a set of orthogonal idempotents whose sum is 1. The results from the last section show that for each $\bar{K}G_n$-module M,

$$M = MA_{n-1}^{(1)} \oplus \Big(\sum_{i=2}^{q^{n-1}} MA_{n-1}^{(i)} \Big) ,$$

as a $\bar{K}G^*_{n-1}$-module.

Given an irreducible $\bar{K}G_n$-module D, the <u>composition multiplicity</u> of D in M means the number of factors isomorphic to D in a composition series of M. The above decomposition of M as a $\bar{K}G_{n-1}$-module, together with the fact that A_{n-1} has order coprime to the characteristic of K shows:

(9.10) The composition multiplicity of the trivial $\bar{K}G_n$-module in arbitrary $\bar{K}G_n$-module M is at most the composition multiplicity of the trivial $\bar{K}G_{n-1}$-module in $MA_{n-1}^{(1)} = ME_1$.

We next investigate more fully the decomposition of M when it is viewed as a $\bar{K}G^*_{n-1}$-module:

9.11 THEOREM. <u>If M is any $\bar{K}G_n$-module, then</u>

(i) $M = ME_1 \oplus \Big(\sum_{r=2}^{n} ME_r \, G^*_{n-1} \Big)$, <u>and</u>

(ii) $\dim M = \sum_{r=1}^{n} (q^{n-1} - 1)(q^{n-2} - 1) \ldots (q^{n-r+1} - 1) \dim (ME_r)$.

<u>Proof</u>: First note that, by Lemma 8.7,

(9.12) All elements of G^*_{n-s} commute with E_r when $1 \leq r \leq s \leq n$.

Now assume that $1 \leq s < n$ and that the following results are true:

(9.13) $\sum_{r=1}^{s} ME_{n-r+1} \, G^*_{s-1} = MA_{n-1}^{(2)} \, A_{n-2}^{(2)} \ldots A_s^{(2)}$, and

$$\dim \Big(\sum_{r=1}^{s} ME_{n-r+1} \, G^*_{s-1} \Big) = \sum_{r=1}^{s} \Big(\prod_{i=r}^{s-1} (q^i - 1) \dim (ME_{n-r+1}) \Big).$$

These results are certainly true when $s = 1$.

Since $G^*_{s-1} \subset H^*_s \subset G^*_s$, we see that the right coset representatives for

G^*_{s-1} in G^*_s can be taken in the form $h_j g_i$, where $h_j \in H^*_s$ and

$$\{g_i \mid 1 \le i \le q^s - 1\}$$

is a set of right coset representatives for H^*_s in G^*_s. Lemma 8.7 shows that

$$(9.14) \quad \{g_i^{-1} A_s^{(2)} g_i \mid 1 \le i \le q^s - 1\} = \{A_s^{(i)} \mid 2 \le i \le q^s\}.$$

Therefore,

$$\sum_{r=1}^{s} ME_{n-r+1} G^*_s = \sum_{r=1}^{s} ME_{n-r+1} G^*_{s-1} (\sum_{j,i} h_j g_i)$$

$$= MA_{n-1}^{(2)} A_{n-2}^{(2)} \dots A_s^{(2)} (\sum_{j,i} h_j g_i)$$

$$= MA_{n-1}^{(2)} A_{n-2}^{(2)} \dots A_s^{(2)} (\sum_i g_i), \text{ by Lemma 8.7,}$$

$$= \sum_{i=2}^{q^s} MA_{n-1}^{(2)} A_{n-2}^{(2)} \dots A_{s+1}^{(2)} A_s^{(i)}, \text{ using (9.14).}$$

However, $ME_{n-s} G^*_s = ME_{n-s}$, since the elements of G^*_s commute with E_{n-s}, and $E_{n-s} = A_{n-1}^{(2)} A_{n-2}^{(2)} \dots A_{s+1}^{(2)} A_s^{(1)}$. Now, $\{A_s^{(i)} \mid 1 \le i \le q^s\}$ is a set of orthogonal idempotents in $\bar{K}G_n$ whose sum is 1, so

$$\sum_{r=1}^{s+1} ME_{n-r+1} G^*_s = ME_{n-s} \oplus \sum_{r=1}^{s} ME_{n-r+1} G^*_s$$

$$= MA_{n-1}^{(2)} A_{n-2}^{(2)} \dots A_{s+1}^{(2)}.$$

This proves that the first part of (9.13) is true when s is replaced by s + 1, and hence (by putting s + 1 = n in the last equation above) we have proved conclusion (i) of the Theorem.

We also obtain:

$$\dim (\sum_{r=1}^{s+1} ME_{n-r+1} G^*_s)$$

$$= \dim (ME_{n-s}) + (q^s - 1) \dim (\sum_{r=1}^{s} ME_{n-r+1} G^*_{s-1})$$

$$= \dim (ME_{n-s}) + \sum_{r=1}^{s} \{(q^s - 1) \prod_{i=r}^{s-1} (q^i - 1) \dim (ME_{n-r+1})\}, \text{ by (9}$$

$$= \sum_{r=1}^{s+1} \prod_{i=r}^{s} (q^i - 1) \dim (ME_{n-r+1}).$$

Therefore, the second part of (9.13) is true when $s + 1$ replaces s. Conclusi (ii) of the Theorem is the result of taking $s = n$ in (9.13). ∎

EXAMPLE. If $n = 4$, then

(i) $M = ME_1G_3^* + ME_2G_3^* + ME_3G_3^* + ME_4G_3^*$, and

(ii) $\dim M = \dim (ME_1) + (q^3 - 1) \dim (ME_2)$

$+ (q^3 - 1)(q^2 - 1) \dim (ME_3) + (q^3 - 1)(q^2 - 1)(q - 1) \dim (ME_4$

.0 THE PERMUTATION MODULE M_λ

Let λ be a composition of n, and let P_λ be the corresponding parabolic subgroup of G_n, described in Section 6.

.0.1 DEFINITIONS.

Let $\bar{P}_\lambda = \sum_{g \in P_\lambda} g \in \bar{K}G_n$, and let $M_\lambda = \bar{P}_\lambda (\bar{K}G_n)$.

Thus, M_λ is a right ideal of $\bar{K}G_n$, and so it is a right $\bar{K}G_n$-module. Every element of M_λ is a $\bar{K}$-linear combination of terms of the form

$$\bar{P}_\lambda \pi u \quad (\pi \in W,\ u \in U^-),$$

by Theorem 7.5, but M_λ may be regarded equally well as the permutation representation of $\bar{K}G_n$ on the set of λ-flags.

.0.2 THEOREM.

Let Γ be a closed subset of Φ^-, and $\pi \in \mathfrak{S}_n$. Define

$$\Gamma_1 = \{(i, j) \in \Gamma \mid \text{row}_{T_\lambda \pi}(i) \leq \text{row}_{T_\lambda \pi}(j)\}$$

$$\Gamma_2 = \{(i, j) \in \Gamma \mid \text{row}_{T_\lambda \pi}(i) > \text{row}_{T_\lambda \pi}(j)\}.$$

Suppose that c and χ_c are as in Definition 9.1. Then

$$\bar{P}_\lambda \pi \sum_{g \in G(\Gamma)} \chi_c(g) g = q^{|\Gamma_1|} \bar{P}_\lambda \pi \sum_{g \in G(\Gamma_2)} \chi_c(g) g \neq 0$$

if $\{2, 3, \ldots, q\} c^{-1} \subseteq \Gamma_2$

$= 0$, *otherwise*.

The cosets $P_\lambda \pi g$ ($g \in G(\Gamma_2)$) are all distinct.

Proof: If $\{2, 3, \ldots, q\} c^{-1} \not\subseteq \Gamma_2$, then $(k, \ell)c > 1$ for some $(k, \ell) \in \Gamma_1$. Then

$$\pi X_{k,\ell} \pi^{-1} \subseteq P_\lambda \quad \text{and} \quad \sum_{\alpha \in \mathbb{F}_q} \chi_{(k,\ell)c}(\alpha) = 0.$$

Using these results, we have

$$\bar{P}_\lambda \pi \sum_{g \in X_{k,\ell}} \chi_c(g)g = \bar{P}_\lambda \pi \sum_{\alpha \in \mathbb{F}_q} \chi_{(k,\ell)c}(\alpha) x_{k,\ell}(\alpha)$$

$$= \sum_{\alpha \in \mathbb{F}_q} (\chi_{(k,\ell)c}(\alpha)\ \bar{P}_\lambda \pi x_{k,\ell}(\alpha))$$

$$= (\sum_{\alpha \in \mathbb{F}_q} \chi_{(k,\ell)c}(\alpha))\ \bar{P}_\lambda \pi$$

$$= 0\ .$$

But

$$\bar{P}_\lambda \pi \sum_{g \in G(\Gamma)} \chi_c(g)g$$

$$= \bar{P}_\lambda \pi\ (\sum_{g \in X_{k,\ell}} \chi_c(g)g).(\text{an element of } \bar{K}G_n), \quad \text{by Theorem 9.2,}$$

$$= 0.$$

If $\{2, 3, \ldots, q\}c^{-1} \subseteq \Gamma_2$, then

$$\bar{P}_\lambda \pi \sum_{g \in G(\Gamma_1)} \chi_c(g)g = \bar{P}_\lambda \pi \sum_{g \in G(\Gamma_1)} g$$

$$= q^{|\Gamma_1|}\ \bar{P}_\lambda \pi.$$

But

$$\sum_{g \in G(\Gamma)} \chi_c(g)g = (\sum_{g \in G(\Gamma_1)} \chi_c(g)g)(\sum_{g \in G(\Gamma_2)} \chi_c(g)g),$$

by Theorem 9.2, so

$$\bar{P}_\lambda \pi \sum_{g \in G(\Gamma)} \chi_c(g)g = q^{|\Gamma_1|}\ \bar{P}_\lambda \pi \sum_{g \in G(\Gamma_2)} \chi_c(g)g\ .$$

Since $\Gamma \subseteq \Phi^-$, Theorem 7.5 shows that the cosets $P_\lambda \pi g$ $(g \in G(\Gamma_2))$ are all distinct, and therefore the above element is non-zero. ■

Next, we investigate the consequences of Theorem 9.11 when $M = M_\lambda$. Recall that $\Gamma(r) = \{(i, j) \mid i > j \leq r\}$, and

$$E_r = \frac{1}{q^{|\Gamma(r)|}} \sum_{g \in G(\Gamma(r))} \chi_{c_r}(g) g = A^{(2)}_{n-1} \ldots A^{(2)}_{n-r+1} A^{(1)}_{n-r},$$

where $(i, j)c_r \neq 1$ only if $i - 1 = j < r$.

Now, G_{n-r} normalizes $G(\Gamma(r))$ and $G(\Gamma(r))G_{n-r}$ is the group of matrices of the form

(10.3)

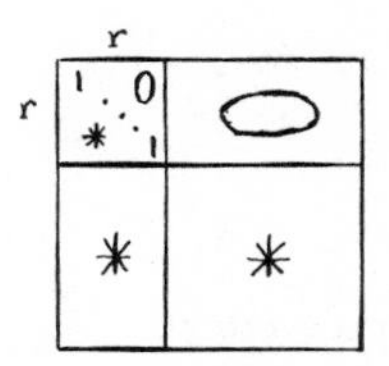

Using (9.12), we see that all elements of $G(\Gamma(r))G_{n-r}$ (and, in particular, all elements of U^-) commute with E_r.

M_λ is spanned by elements of the form $\bar{P}_\lambda \pi u$ ($\pi \in W$, $u \in U^-$), and $\bar{P}_\lambda \pi u E_r = \bar{P}_\lambda \pi E_r u$. Furthermore, an application of Theorem 10.2 shows that $\bar{P}_\lambda \pi E_r \neq 0$ only if

$$\{(j+1, j) \mid 1 \leqq j < r\} \subseteq \{(i, j) \mid \mathrm{row}_{T_\lambda \pi}(i) > \mathrm{row}_{T_\lambda \pi}(j)\}.$$

That is,

(10.4) $\bar{P}_\lambda \pi E_r \neq 0$ only if

$$\mathrm{row}_{T_\lambda \pi}(1) < \mathrm{row}_{T_\lambda \pi}(2) < \ldots < \mathrm{row}_{T_\lambda \pi}(r).$$

For this to be possible, T_λ must have at least r rows. This proves:

10.5 THEOREM. <u>Suppose that λ has precisely h non-zero parts. Then $M_\lambda E_r = 0$ for $r > h$. Thus</u>

(i) $M_\lambda = M_\lambda E_1 \oplus (\sum_{r=2}^{h} M_\lambda E_r G^*_{n-1})$

(ii) $\dim M_\lambda = \sum_{r=1}^{h} (q^{n-1} - 1)(q^{n-2} - 1) \ldots (q^{n-r+1} - 1) \dim (M_\lambda E_r)$.

For the remainder of this Chapter, let $\lambda = (\lambda_1, \lambda_2, \ldots, \lambda_h)$ with each $\lambda_i > 0$. Clearly, this assumption involves no loss. Let the integer satisfy $1 \le r \le h$; we shall examine $M_\lambda E_r$.

10.6 DEFINITION. If $\sigma, \tau \in \mathfrak{S}_n$, write $\sigma \sim \tau$ if $\mathrm{row}_{T_\lambda\sigma}(i) = \mathrm{row}_{T_\lambda\sigma}(i)$ for i with $1 \le i \le r$.

10.7 LEMMA. <u>If $g \in G_{n-r}$ and $\pi \in W$, then there exist $\pi' \in W$ and $u \in U^- \cap$ such that</u>

$$P_\lambda \pi g = P_\lambda \pi' u \quad \underline{\text{and}} \quad \pi \sim \pi'.$$

<u>Proof</u>: There exists $\sigma \in W \cap G_{n-r}$ such that

$$(r + 1)(\pi\sigma)^{-1} > (r + 2)(\pi\sigma)^{-1} > \ldots > n(\pi\sigma)^{-1}.$$

Using the Bruhat Decomposition for G_{n-r}, we may write

$$\sigma^{-1} g = h u_1 \tau u_2 ,$$

where $h \in H$, $u_1 \in U^+$, $\tau \in W$, $u_2 \in U^-$ and $h, u_1, \tau, u_2 \in G_{n-r}$. Then

$$P_\lambda \pi g = P_\lambda \pi\sigma u_1 \tau u_2 .$$

But $P_\lambda \pi\sigma u_1 = P_\lambda \pi\sigma$, by the definition of σ, and $\pi\sigma\tau \sim \pi$, since $\sigma\tau \in W \cap G_{n-}$ Therefore $P_\lambda \pi g = P_\lambda (\pi\sigma\tau) u_2$ gives the expression we seek. ∎

10.8 DEFINITIONS. Assume $1 \le r \le h$.

(i) Let $\mathcal{R}_r = \{R \mid R \subseteq \{1, 2, \ldots, h\} \text{ and } |R| = r\}$.

(ii) For each $R \in \mathcal{R}_r$, let π_R be any permutation such that

$$\mathrm{row}_{T_\lambda \pi_R}(1) < \mathrm{row}_{T_\lambda \pi_R}(2) < \ldots < \mathrm{row}_{T_\lambda \pi_R}(r) ,$$

and $R = \{\mathrm{row}_{T_\lambda \pi_R}(i) \mid 1 \le i \le r\}$.

(iii) For each $R \in \mathcal{R}_r$, define the composition λ_R of $(n - r)$ by

$$\lambda_R = (\lambda_1 - \varepsilon_1, \lambda_2 - \varepsilon_2, \ldots, \lambda_h - \varepsilon_h), \text{ where } \varepsilon_i = 1 \text{ if } i \in R$$

$$\varepsilon_i = 0 \text{ if } i \notin R.$$

By construction, no two elements in $\{\pi_R \mid R \in \mathcal{R}_r\}$ are equivalent, in the sense of 10.6. Hence, by Lemmas 7.3 and 10.7,

(10.9) $\sum_{R \in \mathcal{R}_r} \bar{P}_\lambda \pi_R(\bar{K}G_{n-r})$ is a direct sum.

Now, Theorem 10.2 shows that

(10.10) $$\bar{P}_\lambda \pi_R E_r = \frac{1}{q^{|\Gamma_2|}} \bar{P}_\lambda \pi_R \sum_{g \in G(\Gamma_2)} \chi_c(g) g ,$$

where $\Gamma_1 = \{(i, j) \mid i > j \le r \text{ and } \mathrm{row}_{T_\lambda \pi_R}(i) \le \mathrm{row}_{T_\lambda \pi_R}(j)\}$

$\Gamma_2 = \{(i, j) \mid i > j \le r \text{ and } \mathrm{row}_{T_\lambda \pi_R}(i) > \mathrm{row}_{T_\lambda \pi_R}(j)\}$.

But we cannot have $P_\lambda \pi_R g = P_\lambda \pi_R g'$ with $g \in G(\Gamma_2) \subseteq U^-$, $g' \in G_{n-r}$, unless $g = 1$, by Lemma 10.7 and Theorem 7.5. Therefore,

(10.11) If $\xi \in \bar{K}G_{n-r}$ and $\bar{P}_\lambda \pi_R E_r \xi = 0$, then $\bar{P}_\lambda \pi_R \xi = 0$.

Hence, from (10.9), and the fact that the elements of G_{n-r} commute with E_r, we have

10.12 THEOREM. <u>$\sum_{R \in \mathcal{R}_r} \bar{P}_\lambda \pi_R E_r(\bar{K}G_{n-r})$ is a direct sum. The $\bar{K}G_{n-r}$-modules $\bar{P}_\lambda \pi_R E_r(\bar{K}G_{n-r})$ and $\bar{P}_\lambda \pi_R(\bar{K}G_{n-r})$ are isomorphic</u>.

10.13 THEOREM. <u>For $1 \le r \le h$,</u>

$$M_\lambda E_r = \bigoplus_{R \in \mathcal{R}_r} \bar{P}_\lambda \pi_R E_r(\bar{K}G_{n-r}) .$$

<u>Proof</u>: Let $\pi \in W$, $u \in U^-$. Since $\bar{P}_\lambda \pi u E_r = \bar{P}_\lambda \pi E_r u$, (10.4) shows that $\bar{P}_\lambda \pi u E_r \neq 0$ only if $\pi \sim \pi_R$ for some $R \in \mathcal{R}_r$. But then $\bar{P}_\lambda \pi = \bar{P} \pi_R \sigma$ for some

$\sigma \in W \cap G_{n-r}$. Also, $u = g_1 g_2$ for some $g_1 \in G(\Gamma(r))$, $g_2 \in G_{n-r}$, so

$$\bar{P}_\lambda \pi u E_r = \kappa \bar{P}_\lambda \pi_R E_r g_2 \sigma \quad \text{for some } \kappa \in \bar{K}$$

($g_1 E_r = \kappa E_r$ because $g_1 \in G(\Gamma(r))$, and E_r is the group sum for $G(\Gamma(r))$, weighte[d] by a linear $\bar{K}$-character.) Therefore,

$$M_\lambda E_r = \sum_{R \in \mathcal{R}_r} \bar{P}_\lambda \pi_R E_r (\bar{K}G_{n-r}) ,$$

and (10.12) gives the stated result. ∎

10.14 THEOREM. _If we identify G_{n-r} with $GL_{n-r}(q)$, then $\bar{P}_\lambda \pi_R (\bar{K}G_{n-r})$ and $\bar{P}_{\lambda_R} (\bar{K}G_{n-r})$ are isomorphic $\bar{K}G_{n-r}$-modules._

Proof: We may regard $\bar{P}_{\lambda_R} (\bar{K}G_{n-r})$ as the vector space over $\bar{K}$ spanned by the λ_R-flags

$$\langle e_{r+1}, e_{r+2}, \ldots, e_n \rangle = V_0 \supseteq V_1 \supseteq \ldots \supseteq V_h = 0 ,$$

where $\dim (V_{i-1}/V_i) = (\lambda_R)_i \quad (1 \le i \le h)$.

For each λ_R-flag of this form, define a λ-flag

$$V = V_0^* \supset V_1^* \supset \ldots \supset V_h^* = 0$$

by putting

$$V_i^* = V_i + \langle e_r, e_{r-1}, \ldots, e_m \rangle ,$$

where m is the least integer such that $\text{row}_{T_\lambda \pi_R}(m) > i$. Then $\bar{P}_{\lambda_R} (\bar{K}G_{n-r})$ is isomorphic to the $\bar{K}G_{n-r}$-module consisting of the vector space over $\bar{K}$ spanned by the λ-flags so constructed. But the latter module is, in turn, isomorphi[c] to $\bar{P}_\lambda \pi_R (\bar{K}G_{n-r})$. ∎

Combining our last three Theorems, we have:

10.15 COROLLARY. _For $1 \le r \le h$, there is a $\bar{K}G_{n-r}$-isomorphism,_

$$M_\lambda E_r \cong \bigoplus_{R\in\mathcal{R}_r} M_{\lambda_R}.$$

■

Applying Theorem 10.5, we obtain:

10.16 COROLLARY. <u>If $\lambda = (\lambda_1, \lambda_2, \ldots, \lambda_h)$, all the parts being non-zero, then</u>

(i) $$M_\lambda \cong \Big(\bigoplus_{R\in\mathcal{R}_1} M_{\lambda_R}\Big) \oplus \Big(\sum_{r=2}^{h} M_\lambda E_r G^*_{n-1}\Big),$$ <u>as $\bar{K}G_{n-1}$-modules.</u>

(ii) $$\dim M_\lambda = \sum_{r=1}^{h} (q^{n-1} - 1)(q^{n-2} - 1) \ldots (q^{n-r+1} - 1) \sum_{R\in\mathcal{R}_r} \dim M_{\lambda_R}.$$

■

EXAMPLE. If $\lambda = (\lambda_1, \lambda_2)$ has two non-zero parts, then

$$\dim M_{(\lambda_1,\lambda_2)} = \dim M_{(\lambda_1-1,\lambda_2)} + \dim M_{(\lambda_1,\lambda_2-1)}$$
$$+ (q^{n-1} - 1) \dim M_{(\lambda_1-1,\lambda_2-1)}.$$

■

We remark upon the similarity between Corollary 10.16 and the corresponding results for $\mathfrak{S}_n$. In the case of symmetric groups, M_λ is replaced by the permutation module of $\mathfrak{S}_n$ on the Young subgroup for the composition λ. It is then easy to see (using the same notation for $\bar{K}\mathfrak{S}_n$-modules), that

$$M_\lambda \cong \bigoplus_{R\in\mathcal{R}_1} M_{\lambda_R}, \quad \text{as a } \bar{K}\mathfrak{S}_{n-1}\text{-module.}$$

Thus, there is no module corresponding to the second term in 10.16(i). But, on the other hand, the second term has dimension divisible by $q - 1$, so when we "put $q = 1$, the second term disappears"!

To conclude this chapter, we give an alternative formula for $\dim M_\lambda$, and derive a numerical result concerning this dimension.

10.17 DEFINITION. (cf. Definition 2.5) Let k be an integer, and define

$\{k\} = [k][k-1] \ldots [1] \quad \text{if } k \geq 1$

$\{0\} = 1$

$1/\{k\} = 0 \quad \text{if } k < 0.$

Since M_λ is the permutation module of G_n on P_λ, $\dim M_\lambda = |G_n : P_\lambda|$. Hence

$$(10.18) \quad \dim M_\lambda = \frac{\{n\}}{\{\lambda_1\}\{\lambda_2\} \ldots \{\lambda_h\}} .$$

We note that it is not obvious that this agrees with 10.16(ii).

10.19 DEFINITION. If μ is a <u>partition</u> of n, let

$$\dim(\mu) = \{n\} \det \left(\frac{1}{\{\mu_i + j - i\}} \right)$$

EXAMPLE. If $\mu = (4, 3, 1)$, then

$$\dim(\mu) = \{8\} . \begin{vmatrix} \frac{1}{\{4\}} & \frac{1}{\{5\}} & \frac{1}{\{6\}} \\ \frac{1}{\{2\}} & \frac{1}{\{3\}} & \frac{1}{\{4\}} \\ 0 & \frac{1}{\{0\}} & \frac{1}{\{1\}} \end{vmatrix}$$

The reason for the notation is that $\dim(\mu)$ turns out to be the dimension of the module S_μ which we shall construct in the next chapter. The $\bar{K}G_n$-module S_μ is analogous to the Specht module for $\mathfrak{S}_n$.

10.20 DEFINITION. (cf. 10.8(i)). Assume that $\mu = (\mu_1, \mu_2, \ldots, \mu_h)$ is a <u>partition</u> of n, with $\mu_h > 0$, and $1 \leq r \leq h$. Let $\mathcal{R}_r^*$ be the set of subsets R of $\{1, 2, \ldots, h\}$ which have the properties that $|R| = r$ and if $b \in R$ and $\mu_b = \mu_{b+1}$, then $b + 1 \in R$.

Thus, an element $R \in \mathcal{R}_r$ lies in $\mathcal{R}_r^*$ if and only if μ_R is a partition of $n - r$.

10.21 THEOREM. Assume that μ is a partition of n. Then

$$\dim(\mu) = \sum_{r=1}^{h} \sum_{R \in \mathcal{R}_r^*} (q^{n-1} - 1)(q^{n-2} - 1) \ldots (q^{n-r+1} - 1) \dim(\mu_R).$$

Proof: All the terms in the expansion of the determinant for dim (μ) have the form

$$\frac{\{n\}}{\{a_1\}\{a_2\} \ldots \{a_h\}} \quad \text{where } a_1 + \ldots + a_h = n.$$

Hence, by Corollary 10.16 and (10.18),

$$\dim(\mu) = \sum_{r=1}^{h} \sum_{R \in \mathcal{R}_r} (q^{n-1} - 1) \ldots (q^{n-r+1} - 1) \dim(\mu_R).$$

But if $b \in R$ and $\mu_b = \mu_{b+1}$ and $b + 1 \notin R$, then

$$\mu_R = (\ldots, \mu_b - 1, \mu_b, \ldots)$$

and dim $(\mu_R) = 0$, since rows b, b + 1 of the determinant are equal in this case. The stated results follows. ■

EXAMPLE. If $\mu = (4, 2, 2)$, then

$$[\mu] = \begin{matrix} X & X & X & X \\ X & X & & \\ X & X & & \end{matrix} \quad .$$

$$\begin{aligned} \text{Dim } M_\mu &= \dim M_{(3,2,2)} + \dim M_{(4,1,2)} + \dim M_{(4,2,1)} \\ &+ (q^7 - 1)(\dim M_{(3,1,2)} + \dim M_{(3,2,1)} + \dim M_{(4,1,1)}) \\ &+ (q^7 - 1)(q^6 - 1) \dim M_{(3,1,1)}. \end{aligned}$$

$$\begin{aligned} \dim(\mu) &= \dim(3, 2, 2) + \dim(4, 2, 1) + (q^7 - 1)(\dim(3, 2, 1) + \dim(4, 1, 1)) \\ &+ (q^7 - 1)(q^6 - 1) \dim(3, 1, 1). \end{aligned}$$

11 THE SUBMODULE THEOREM

We are now in a position to construct an irreducible $\bar{K}G_n$-module for each composition λ of n.

11.1 DEFINITION. Let $< , >_\lambda$ be the unique bilinear form on M_λ such that

$$\langle \bar{P}_\lambda g_1, \bar{P}_\lambda g_2 \rangle_\lambda = 1 \quad \text{if } P_\lambda g_1 = P_\lambda g_2$$
$$= 0 \quad \text{if } P_\lambda g_1 \neq P_\lambda g_2 \qquad (g_1, g_2 \in G_n).$$

This form is clearly non-singular, symmetric, and G_n-invariant. If I is any subset of M_λ, we define $I^\perp$ by

$$I^\perp = \{m' \in M_\lambda \mid \langle m, m' \rangle_\lambda = 0 \quad \text{for all } m \in I\}.$$

Since the bilinear form is G_n-invariant, $I^\perp$ is a $\bar{K}G_n$-submodule of M_λ if I is a $\bar{K}G_n$-submodule of M_λ.

Our plan is to apply the following general theorem:

11.2 THEOREM. Let G be any group, K an arbitrary field and suppose that M is any right KG-module, endowed with a G-invariant non-singular bilinear form $< , >$. Assume that there exists an idempotent $E = \sum_i a_i g_i \in KG$ ($a_i \in K$, $g_i \in G$) with the properties (a), (b), (c) listed below:

(a) ME is a 1-dimensional space, ME = KmE, say ($m \in M$).

(b) If $\bar{E} = \sum_i a_i g_i^{-1}$, then $mE(KG) = m\bar{E}(KG)$.

(c) $\langle mE, m \rangle \neq 0$.

Define S = mE(KG). Then each of the following results is true:

(i) For every KG-submodule I of M, either $I \supseteq S$ or $I \subseteq S^\perp$.

(ii) $S \nsubseteq S^\perp$.

(iii) If I is any KG-submodule of M and θ is a KG-homomorphism from S into M/I, then for some $\kappa \in K$, $(mE)\theta = \kappa mE + I$.

(iv) $S/(S \cap S^\perp)$ is a self-dual absolutely irreducible KG-module.

(v) <u>The composition multiplicity of $S/(S \cap S^{\perp})$ in M is one.</u>

<u>Proof</u>: (i) If, for some $x \in I$, $xE \neq 0$, then $mE \in I$, by hypothesis (a), and $I \supseteq S$. On the other hand, if $xE = 0$ for all $x \in I$, then for all $x \in I$, $g \in G$,

$$0 = \langle xg^{-1}E, m\rangle = \langle x, m\bar{E}g\rangle,$$

since $\langle\ ,\ \rangle$ is G-invariant. Then $I \subseteq S^{\perp}$, by hypothesis (b).

(ii) mE, $m\bar{E} \in S$, by hypothesis (b). Also,

$$\langle mE, m\bar{E}\rangle = \langle mE^2, m\rangle = \langle mE, m\rangle \neq 0,$$

by hypothesis (c), so $S \nsubseteq S^{\perp}$.

(iii) If θ is a KG-homomorphism from S into M/I, then

$$(mE)\theta = (mE^2)\theta = (mE)\theta E = \kappa(mE) + I,$$

for some $\kappa \in K$, by hypothesis (a).

(iv) and (v): $S/(S \cap S^{\perp})$ is an irreducible KG-module, in view of conclusions (i) and (ii).

$(S \cap S^{\perp})^{\perp} = S + S^{\perp}$, so the dual of $S/(S \cap S^{\perp})$ is isomorphic to $(S + S^{\perp})/S^{\perp}$, which in turn is isomorphic to $S/(S \cap S^{\perp})$, by the Second Isomorphism Theorem. Therefore, $S/(S \cap S^{\perp})$ is self-dual.

If I is a KG-submodule of M, and $\bar{\theta}$ is a KG-homomorphism from $S/(S \cap S^{\perp})$ into M/I, then by composing the maps

$$S \xrightarrow[\text{canon.}]{} S/(S \cap S^{\perp}) \xrightarrow[\bar{\theta}]{} M/I,$$

we get a KG-homomorphism θ from S into M/I. Hence

$$\bar{\theta}: mE + (S \cap S^{\perp}) \mapsto \kappa mE + I$$

for some $\kappa \in K$, by conclusion (iii). Now conclusion (v) follows at once. Taking $I = S \cap S^{\perp}$, we have also proved that every KG-endomorphism of $S/(S \cap S^{\perp})$ is just multiplication by an element of K. Therefore, $S/(S \cap S^{\perp})$ is absolutely irreducible (see, for example, Dornhoff [D], p. 20). ∎

It is noteworthy that although the conclusions of this Theorem hold when $G = \mathfrak{S}_n$ (taking M to be the permutation module in a Young subgroup - see James [J_1]), we do not know an appropriate idempotent E, except for certain fields K. A similar theorem is true for Weyl modules; see Theorem 8.3.2 in James and Kerber [JK]. In view of the results we have already proved, there is no difficulty in devising a suitable idempotent E to apply Theorem 11.2 in the case where $G = G_n$ and $M = M_\lambda$.

Recall that T_λ is the λ-tableau which has 1, 2, ..., n in order along successive rows.

11.3 DEFINITION. Let π_λ be the permutation which sends T_λ to the λ-tableau which has 1, 2, ..., n in order reading from top to bottom in successive columns.

EXAMPLE. If $\lambda = (3, 2, 4)$, then

$$T_\lambda = \begin{matrix} 1 & 2 & 3 & \\ 4 & 5 & & \\ 6 & 7 & 8 & 9 \end{matrix} \qquad T_\lambda\pi_\lambda = \begin{matrix} 1 & 4 & 7 & \\ 2 & 5 & & \\ 3 & 6 & 8 & 9 \end{matrix} \quad .$$

Inspired by the definition of the idempotent E_r which was used in the last two sections, we give:

11.4 DEFINITION (cf. 9.1 and 9.4).

Let c be the function from Φ^- to $\{1, 2, ..., q\}$, defined by

$(i, j)c = 2$, if $i = j + 1$ and $\mathrm{col}_{T_\lambda\pi_\lambda}(i) = \mathrm{col}_{T_\lambda\pi_\lambda}(j)$

$(i, j)c = 1$, otherwise.

Let

$$E_\lambda = \frac{1}{|U^-|} \sum_{g\in U^-} \chi_c(g) g \ .$$

Again, the exact nature of c is unimportant. For (9.6) tells us that when c' is any other function from Φ^- to $\{1, 2, \ldots, q\}$ such that

$$(i, j)c' > 1 \quad \text{if } i = j + 1 \text{ and } \mathrm{col}_{T_\lambda\pi_\lambda}(i) = \mathrm{col}_{T_\lambda\pi_\lambda}(j)$$

$$(i, j)c' = 1, \quad \text{otherwise,}$$

we have, for some $h \in H$,

(11.5) $$h^{-1}E_\lambda h = \frac{1}{|U^-|} \sum_{g\in U^-} \chi_{c'}(g)g.$$

In particular,

(11.6) Let $\bar{E} = \dfrac{1}{|U^-|} \sum_{g\in U^-} \chi_c^{-1}(g)g.$

Then $\bar{E}_\lambda = h^{-1}E_\lambda h$ for some $h \in H$.

EXAMPLES

(i) If $\lambda = (3, 2, 4)$, then $E_\lambda = A_8^{(2)}A_7^{(2)}A_6^{(1)}A_5^{(2)}A_4^{(2)}A_3^{(1)}A_2^{(2)}A_1^{(1)}$.

(ii) If $\lambda = (3, 2)$, then $T_\lambda\pi_\lambda = \begin{matrix} 1 & 3 & 5 \\ 2 & 4 & \end{matrix}$, and

$$E_\lambda = \frac{1}{q^{10}} \sum \chi_2(\alpha)\chi_2(\beta) \begin{pmatrix} 1 & & & & \\ \alpha & 1 & & 0 & \\ * & * & 1 & & \\ * & * & \beta & 1 & \\ * & * & * & * & 1 \end{pmatrix},$$

the sum being over all lower unitriangular matrices.

(iii) If $\lambda = (1, 1, 1)$, then $T_\lambda\pi_\lambda = \begin{matrix} 1 \\ 2 \\ 3 \end{matrix}$, and

$$E_\lambda = \frac{1}{q^3} \sum_{\alpha,\beta,\gamma\in\mathbb{F}_q} \chi_2(\alpha)\chi_2(\beta) \begin{pmatrix} 1 & 0 & 0 \\ \alpha & 1 & 0 \\ \gamma & \beta & 1 \end{pmatrix}.$$

The crucial properties of the idempotent E_λ are given by:

11.7 THEOREM. Let λ, μ be compositions of n. Then

(i) $M_\mu E_\lambda = 0$ unless $\lambda \trianglerighteq \mu$

(ii) $M_\lambda E_\lambda = \bar{K}\bar{P}_\lambda \pi_\lambda E_\lambda \neq 0$.

<u>Proof</u>: Every element of M_μ is a linear combination of terms of the form $\bar{P}_\mu \pi u$ ($\pi \in W$, $u \in U^-$). But

$$uE_\lambda = \chi_c^{-1}(u)E_\lambda,$$

so $\bar{P}_\mu \pi u E_\lambda \neq 0$ only if $\bar{P}_\mu \pi E_\lambda \neq 0$. By Theorem 10.2, $\bar{P}_\mu \pi E_\lambda \neq 0$ if and only if

$$(11.8)\quad \{(j+1, j) \mid \mathrm{col}_{T_\lambda \pi_\lambda}(j+1) = \mathrm{col}_{T_\lambda \pi_\lambda}(j)\}$$
$$\subseteq \{(i, j) \mid i > j \text{ and } \mathrm{row}_{T_\mu \pi}(i) > \mathrm{row}_{T_\mu \pi}(j)\}.$$

This happens only if the numbers from each column of $T_\lambda \pi_\lambda$ appear in different rows of $T_\mu \pi$. Hence $\bar{P}_\mu \pi E_\lambda = 0$ unless $\lambda \trianglerighteq \mu$ (see Definition 4.1). This proves part (i) of the theorem.

If $\lambda = \mu$, (11.8) holds only for those permutations π for which

$$\mathrm{row}_{T_\lambda \pi_\lambda}(i) = \mathrm{row}_{T_\lambda \pi}(i), \quad \text{for all } i.$$

But then $P_\lambda \pi = P_\lambda \pi_\lambda$, so $\bar{P}_\lambda \pi u E_\lambda$ is a multiple of $\bar{P}_\lambda \pi_\lambda E_\lambda \neq 0$. ■

Applying Theorem 10.2 again, we get

$$(11.9)\quad \bar{P}_\lambda \pi_\lambda E_\lambda = \frac{1}{q^{|\Gamma|}} \bar{P}_\lambda \pi_\lambda \sum_{g \in G(\Gamma)} \chi_c(g) g,$$

where $\Gamma = \{(i, j) \mid i > j \text{ and } \mathrm{row}_{T_\lambda \pi_\lambda}(i) > \mathrm{row}_{T_\lambda \pi_\lambda}(j)\}$.

Hence (noting the last sentence of Theorem 10.2):

$$(11.10)\quad \langle \bar{P}_\lambda \pi_\lambda E_\lambda, \bar{P}_\lambda \pi_\lambda \rangle_\lambda = \frac{1}{q^{|\Gamma|}} \neq 0.$$

11.11 DEFINITIONS. For each composition λ of n, let S_λ, D_λ be the following $\bar{K}G_n$-modules:

$$S_\lambda = \bar{P}_\lambda \pi_\lambda E_\lambda (\bar{K}G_n)$$

$$D_\lambda = S_\lambda/(S_\lambda \cap S_\lambda^\perp).$$

Theorem 11.7(ii), (11.6) and (11.10) prove that the hypotheses of Theorem 11.2 hold when $G = G_n$, $M = M_\lambda$, $S = S_\lambda$. We now list the information we have proved about the modules S_λ, D_λ:

(11.12) (i) Let c be any function from Φ^- to $\{1, 2, \ldots, q\}$ such that

$(i, j)c > 1$ if $i = j + 1$ and $\mathrm{col}_{T_\lambda\pi_\lambda}(i) = \mathrm{col}_{T_\lambda\pi_\lambda}(j)$

$(i, j)c = 1$, otherwise.

Then

$$S_\lambda = \bar{P}_\lambda\pi_\lambda \sum_{g\in U^-} \chi_c(g)g\ (\bar{K}G_n)$$

$$= \bar{P}_\lambda\pi_\lambda \sum_{g\in G(\Gamma)} \chi_c(g)g\ (\bar{K}G_n)$$

where $\Gamma = \{(i, j) \mid i > j \text{ and } \mathrm{row}_{T_\lambda\pi_\lambda}(i) > \mathrm{row}_{T_\lambda\pi_\lambda}(j)\}$.

The cosets $P_\lambda\pi_\lambda g\ (g \in G(\Gamma))$ are all distinct.

(ii) THE SUBMODULE THEOREM. _For every submodule I of M_λ, either $I \supseteq S_\lambda$ or $I \subseteq S_\lambda^\perp$._

(iii) $D_\lambda = S_\lambda/(S_\lambda \cap S_\lambda^\perp)$ is a self-dual, absolutely irreducible $\bar{K}G_n$-module.

(iv) The composition multiplicity of D_λ in M_λ, and in S_λ, is one. ■

Theorem 11.7(i) gives information concerning two compositions λ and μ of n:

11.13 LEMMA. _If D_λ is a composition factor of M_μ, then $\lambda \trianglerighteq \mu$._

Proof: Suppose that I is a $\bar{K}G_n$-submodule of M_μ. Then, as in the proof of Theorem 11.2(v), any non-zero $\bar{K}G_n$-homomorphism $\bar{\theta}$ from D_λ into M_μ/I gives rise to a non-zero $\bar{K}G_n$-homomorphism θ from S_λ into M_μ/I. But

$$\bar{P}_\lambda \pi_\lambda E_\lambda \theta = \bar{P}_\lambda \pi_\lambda E_\lambda^2 \theta = \bar{P}_\lambda \pi_\lambda E_\lambda \theta E_\lambda ,$$

which is zero unless $\lambda \trianglerighteq \mu$, by Theorem 11.7(i). ■

11.14 COROLLARY

(i) If D_μ is a composition factor of S_λ, then $\mu \trianglerighteq \lambda$.

(ii) $D_\lambda \cong D_\mu$ only if μ is obtained from λ by rearranging the parts.

(iii) $S_\lambda \subseteq \bigcap_\theta \mathrm{Ker}\theta$, the intersection being over all $\bar{K}G_n$-homomorphisms θ which map M_λ into some M_μ with $\mu \triangleright \lambda$. ■

It is useful to give a generator of S_λ which consists of a linear combination of distinct terms $\bar{P}_\lambda \pi_\lambda u$ $(u \in U^-)$, each of which has coefficient ± 1. Then the coefficients will belong to K, whereas the generator described in 11.12(i) has coefficients which are roots of unity, which might belong to $\bar{K} \setminus K$.

Let a, a + 1 be integers which lie in the same column of $T_\lambda \pi_\lambda$. Applying Theorem 9.2, we know that

$$\bar{P}_\lambda \pi_\lambda Y \sum_{\alpha \in \mathbb{F}_q} \chi_{(a+1,a)c}(\alpha) x_{a+1,a}(\alpha)$$

generates S_λ, where

$$Y = \prod_{\substack{(i,j)\in\Gamma \\ (i,j)\neq(a+1,a)}} \sum_{\alpha\in\mathbb{F}_q} \chi_{(i,j)c}(\alpha) x_{ij}(\alpha)$$

and Γ is as in 11.12(i).

Now allow c to vary over the q - 1 functions for which $(a + 1, a)c \in \{2, 3, \ldots, q\}$ but which agree with the chosen function c on all other elements $(i, j) \in \Phi^-$. Then

$$\sum_{\alpha \in \mathbb{F}_q} \chi_{(a+1,a)c}(\alpha) x_{a+1,a}(\alpha)$$

varies over a set of generators of the augmentation ideal of the group

$x_{a+1,a}$. Hence S_λ is generated by

$$\bar{P}_\lambda \pi_\lambda Y\,(x_{a+1,a}(0) - x_{a+1,a}(1)).$$

Repeating this process, we deduce that S_λ is generated by

$$(11.15)\quad \bar{P}_\lambda \pi_\lambda \prod_{(i,j)\in\Phi_2} \Big(\sum_{\alpha\in\mathbb{F}_q} x_{ij}(\alpha)\Big) \prod_{(i,j)\in\Phi_1} (x_{ij}(0) - x_{ij}(1)),$$

where $\Phi_1 = \{(j+1, j) \mid \mathrm{col}_{T_\lambda \pi_\lambda}(j+1) = \mathrm{col}_{T_\lambda \pi_\lambda}(j)\}$

and $\Phi_2 = \{(i, j) \in \Phi^- \setminus \Phi_1 \mid \mathrm{row}_{T_\lambda \pi_\lambda}(i) > \mathrm{row}_{T_\lambda \pi_\lambda}(j)\}$.

We have now achieved our aim of producing a generator of S_λ which is a linear combination, with coefficients ± 1, of distinct terms $\bar{P}_\lambda \pi_\lambda u$ $(u \in U^-)$. If we define $S_{\lambda,K}$ to be the right ideal of KG_n generated by the element (11.15), then $S_\lambda = S_{\lambda,K} \otimes \bar{K}$. Also, $\dim(S_{\lambda,K} \cap S_{\lambda,K}^\perp) = \dim(S_\lambda \cap S_\lambda^\perp)$, since each is just the rank of the Gram matrix for the bilinear form $\langle\ ,\ \rangle_\lambda$ with respect to a basis of $S_{\lambda,K}$. Therefore,

$$D_{\lambda,K} = S_{\lambda,K}/(S_{\lambda,K} \cap S_{\lambda,K}^\perp)$$

is a self-dual absolutely irreducible KG_n-module. Although we proved this result, with the generator (11.15) of $S_{\lambda,K}$ in [J_9], the proof given here, using the extension field $\bar{K}$, is superior, and the work in later chapters is greatly eased by the fact that our ground field is $\bar{K}$.

11.16 THEOREM (The Ordinary Irreducible Unipotent Representations of G_n).

_$S_{\lambda,\mathbb{Q}}$ is a self-dual, absolutely irreducible $\mathbb{Q}G_n$-module. If λ and μ are distinct partitions of n, then $S_{\lambda,\mathbb{Q}}$ is not isomorphic to $S_{\mu,\mathbb{Q}}$._

Proof: When $K = \mathbb{Q}$, $\langle\ ,\ \rangle_\lambda$ is an inner product. Therefore, $D_{\lambda,\mathbb{Q}} = S_{\lambda,\mathbb{Q}}$. The result now follows from 11.12(iii) and Corollary 11.14(ii). ■

We now illustrate generators for the module S_λ.

11.17 EXAMPLES

(i) If $\lambda = (n)$, then $M_\lambda = S_\lambda = D_\lambda$ = the trivial $\bar{K}G_n$-module.

(ii) If $\lambda = (1, 1, 1)$, then S_λ is generated by

$$\bar{P}_\lambda \sum_{\alpha,\beta,\gamma\in\mathbb{F}_q} \chi_2(\alpha)\chi_2(\alpha) \begin{pmatrix} 1 & 0 & 0 \\ \alpha & 1 & 0 \\ \gamma & \beta & 1 \end{pmatrix}.$$

It is also generated by

$$\bar{P}_\lambda \sum_{\gamma\in\mathbb{F}_q} \left\{ \begin{pmatrix} 1 & 0 & 0 \\ 0 & 1 & 0 \\ \gamma & 0 & 1 \end{pmatrix} - \begin{pmatrix} 1 & 0 & 0 \\ 1 & 1 & 0 \\ \gamma & 0 & 1 \end{pmatrix} - \begin{pmatrix} 1 & 0 & 0 \\ 0 & 1 & 0 \\ \gamma & 1 & 1 \end{pmatrix} + \begin{pmatrix} 1 & 0 & 0 \\ 1 & 1 & 0 \\ \gamma & 1 & 1 \end{pmatrix} \right\}.$$

(iii) If $\lambda = (3, 2)$, then S_λ is generated by

$$\bar{P}_\lambda \sum_{\alpha,\beta,\gamma\in\mathbb{F}_q} \chi_2(\alpha)\chi_2(\beta) \left(\begin{array}{ccccc} 1 & 0 & 0 & 0 & 0 \\ 0 & 0 & 1 & 0 & 0 \\ 0 & 0 & 0 & 0 & 1 \\ \hline \alpha & 1 & 0 & 0 & 0 \\ \gamma & 0 & \beta & 1 & 0 \end{array}\right)$$

and by

$$\bar{P}_\lambda \sum_{\gamma\in\mathbb{F}_q} \left\{ \left(\begin{array}{ccccc} 1 & 0 & 0 & 0 & 0 \\ 0 & 0 & 1 & 0 & 0 \\ 0 & 0 & 0 & 0 & 1 \\ \hline 0 & 1 & 0 & 0 & 0 \\ \gamma & 0 & 0 & 1 & 0 \end{array}\right) - \left(\begin{array}{ccccc} 1 & 0 & 0 & 0 & 0 \\ 0 & 0 & 1 & 0 & 0 \\ 0 & 0 & 0 & 0 & 1 \\ \hline 1 & 1 & 0 & 0 & 0 \\ \gamma & 0 & 0 & 1 & 0 \end{array}\right) - \left(\begin{array}{ccccc} 1 & 0 & 0 & 0 & 0 \\ 0 & 0 & 1 & 0 & 0 \\ 0 & 0 & 0 & 0 & 1 \\ \hline 0 & 1 & 0 & 0 & 0 \\ \gamma & 0 & 1 & 1 & 0 \end{array}\right) + \left(\begin{array}{ccccc} 1 & 0 & 0 & 0 & 0 \\ 0 & 0 & 1 & 0 & 0 \\ 0 & 0 & 0 & 0 & 1 \\ \hline 1 & 1 & 0 & 0 & 0 \\ \gamma & 0 & 1 & 1 & 0 \end{array}\right) \right\}.$$

(iv) If $\lambda = (3, 2, 4)$, then

$$T_\lambda = \begin{matrix} 1 & 2 & 3 & \\ 4 & 5 & & \\ 6 & 7 & 8 & 9 \end{matrix} \qquad T_\lambda\pi_\lambda = \begin{matrix} 1 & 4 & 7 & \\ 2 & 5 & & \\ 3 & 6 & 8 & 9 \end{matrix}.$$

S_λ is generated by

$\bar{P}_\lambda \sum \pm$

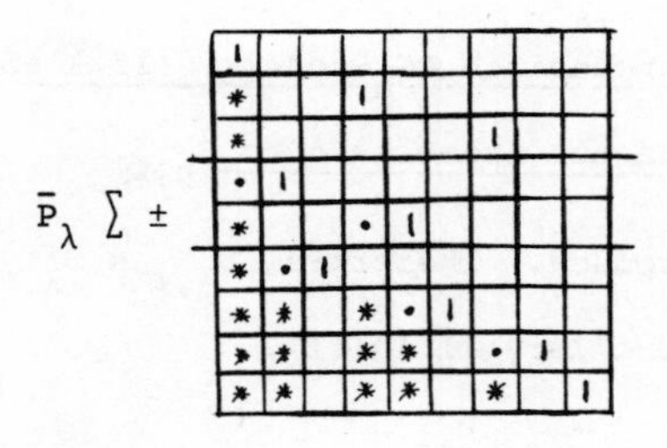

where $\sum$ is the sum over all matrices obtained by replacing each * by an arbitrary element of $\mathbb{F}_q$, and each . by 0 or 1, the sign being + if an even number of .'s are replaced by 1 and −, otherwise.

(v) If $\lambda = (n - m, m)$ with $n - m \geq m$, then S_λ is generated by

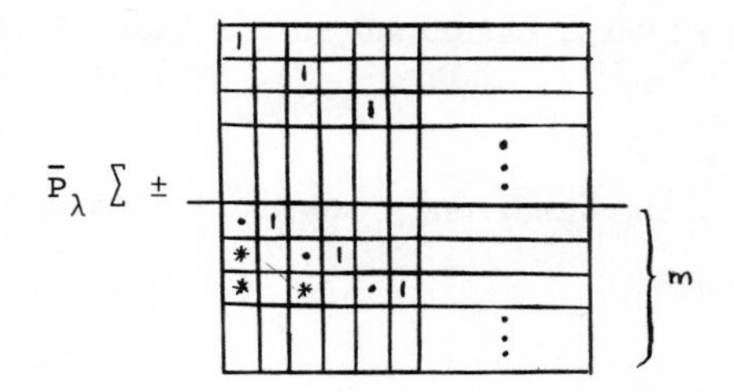

using the same conventions as in Example (iv).

If we regard M_λ as the vector space over $\bar{K}$ spanned by the m-dimensional subspaces of V, then the generator described above is

$$\sum_{\substack{\varepsilon_{ij}=0 \text{ or } 1 \\ \alpha_{ij} \in \mathbb{F}_q}} \pm \langle \varepsilon_{11}e_1 + e_2, \\ \alpha_{21}e_1 + \varepsilon_{23}e_3 + e_4, \\ \alpha_{31}e_1 + \alpha_{33}e_3 + \varepsilon_{35}e_5 + e_6, \\ \vdots \\ \alpha_{m1}e_1 + \alpha_{m3}e_3 + \ldots + \alpha_{m,2m-3}e_{2m-3} + \varepsilon_{m,2m-1}e_{2m-1} + e_{2m}\rangle,$$

where the sign is +/− according to whether the number of ε_{ij}'s equal to 1 is even/odd.

Thus, if $m = 2$ and $n \geq 4$, as in Example (iii), we get

$$\sum_{\alpha_{21} \in \mathbb{F}_q} (\langle e_2, \alpha_{21}e_1 + e_4\rangle - \langle e_1 + e_2, \alpha_{21}e_1 + e_4\rangle \\ - \langle e_2, \alpha_{21}e_1 + e_3 + e_4\rangle + \langle e_1 + e_2, \alpha_{21}e_1 + e_3 + e_4\rangle). \blacksquare$$

Several questions arise from the work in this section; later all the questions listed below will be answered in the affirmative:

(11.18) (i) Suppose that μ is obtained from λ by rearranging the parts. Do we have $M_\lambda \cong M_\mu$, $S_\lambda \cong S_\mu$, $D_\lambda \cong D_\mu$?

(ii) Is every composition factor of S_λ (and M_λ) of the form D_μ with $\mu \trianglerighteq \lambda$?

(iii) Is dim S_λ independent of K?

(iv) If K is the field of p_K elements (p_K a prime not dividing q), is $S_{\lambda,K}$ a module obtained by "reducing modulo p_K" the module $S_{\lambda,\mathbb{Q}}$?

(v) Is the inclusion, $S_\lambda \subseteq \bigcap_\theta \mathrm{Ker}\theta$, described in Corollary 11.14(iii actually equality?

The last question is, in fact, the vital one. We shall call the equality

$$S_\lambda = \bigcap_\theta \mathrm{Ker}\theta$$

"the Kernel Intersection Theorem", and the aim of the next few sections is to prove this result.

12 A LOWER BOUND FOR THE DIMENSION OF S_μ

In this section we shall construct a certain number of linearly independent elements of S_μ when μ is a partition of n. This number will later be shown to be equal to the dimension of S_λ. Since we shall prove that $S_\mu \cong S_\lambda$ when μ is obtained from λ be rearranging the order of the parts, there is no great loss in restricting μ to be a partition, so we assume throughout this chapter that

$$\mu = (\mu_1, \mu_2, \ldots, \mu_h),$$

with $\mu_1 \geq \mu_2 \geq \ldots \geq \mu_h > 0$.

Recall from 10.20 that when $1 \leq r \leq h$, $\mathcal{R}_r^*$ is the set of those subsets R of $\{1, 2, \ldots, h\}$ which have cardinality r, and for which μ_R is a partition of $n - r$.

12.1 DEFINITION. Given $R \in \mathcal{R}_r^*$, construct a μ-tableau T_R according to the following recipe:

(i) For each $b \in R$, place at the end of row b one of the numbers in $\{1, 2, \ldots, r\}$; do this in such a way that

$$\text{row}_{T_R}(1) < \text{row}_{T_R}(2) < \ldots < \text{row}_{T_R}(r).$$

(ii) Complete T_R by inserting the numbers $r + 1, r + 2, \ldots, n$ in order down successive columns in the remaining places.

(iii) Define the permutation π_R by $T_\mu \pi_R = T_R$.

These conditions determine T_R and π_R uniquely, and π_R satisfies 10.8(ii).

EXAMPLE. If $\mu = (3, 3, 2, 1)$ and $R = \{1, 2, 4\}$, then

$$T_\mu = \begin{matrix} 1 & 2 & 3 \\ 4 & 5 & 6 \\ 7 & 8 & \\ 9 & & \end{matrix} \qquad T_\mu \pi_\mu = \begin{matrix} 1 & 5 & 8 \\ 2 & 6 & 9 \\ 3 & 7 & \\ 4 & & \end{matrix} \qquad T_\mu \pi_R = \begin{matrix} 4 & 7 & 1 \\ 5 & 8 & 2 \\ 6 & 9 & \\ 3 & & \end{matrix},$$

$$\pi_R = (1\ 4\ 5\ 8\ 9\ 3)(2\ 7\ 6).$$

12.2 LEMMA. Assume that $i > j$, $i\pi_\mu > j\pi_\mu$, and $R \in \mathcal{R}_r^*$. Then one of the following holds:

(i) $i\pi_R \le r < j\pi_R$

(ii) $r < j\pi_R < i\pi_R$

(iii) $j\pi_R < i\pi_R \le r$. In this case, i and j belong to the same column of T_μ, and $i\pi_\mu = j\pi_\mu + 1$ if and only if $i\pi_R = j\pi_R + 1$.

Proof: Since $i > j$, $\text{row}_{T_\mu}(i) \ge \text{row}_{T_\mu}(j)$.

Now, if $j\pi_R \le r$, then j lies at the end of a row of T_μ, say at the end of row b. Since $i\pi_\mu > j\pi_\mu$, we conclude that $i\pi_\mu$ lies at the end of row c of $T_\mu\pi_\mu$ where

$b < c$, and

$\mu_b = \mu_{b+1} = \ldots = \mu_c$.

Therefore, $j\pi_R < i\pi_R \le r$, and the remainder of statement (iii) follows.

If $r < j\pi_R$, then we do not have $r < i\pi_R < j\pi_R$; for if this were the case, then the construction of T_R would imply that $i\pi_\mu < j\pi_\mu$, a contradiction. Therefore, if $r < j\pi_R$, either conclusion (i) or conclusion (ii) holds. ■

Compare the next result with Corollary 10.15 (cf. also 16.9).

12.3 THEOREM. For $1 \le r \le h$,

$$\dim (S_\mu E_r) \ge \sum_{R \in \mathcal{R}_r^*} \dim S_{\mu_R}.$$

Proof: S_μ is generated by

$$\bar{P}_\mu \pi_\mu \sum_{g \in G(\Gamma)} \chi_c(g) g,$$

where $\Gamma = \{(i, j) \mid i > j \text{ and } \text{row}_{T_\mu\pi_\mu}(i) > \text{row}_{T_\mu\pi_\mu}(j)\}$,

and c is the function from Φ^- to $\{1, 2, \ldots, q\}$ such that

$(i, j)c = 2$, if $i = j + 1$ and $\mathrm{col}_{T_\mu \pi_\mu}(i) = \mathrm{col}_{T_\mu \pi_\mu}(j)$

$(i, j)c = 1$, otherwise,

(see 11.12(i)).

Suppose that $R \in \mathcal{R}_r^*$, and apply $\pi_\mu^{-1}\pi_R$ to this generator:

$$\bar{P}_\mu \pi_R \sum_{g \in G(\Gamma\pi_\mu^{-1}\pi_R)} \chi_{\pi_R^{-1}\pi_\mu c}(g)g \in S_\mu.$$

(Here we have used Theorem 9.2 and 5.10(i); note that $\pi_R^{-1}\pi_\mu c$ is a function from $\Gamma\pi_\mu^{-1}\pi_R$ to $\{1, 2, \ldots, q\}$).

Now,

$$\Gamma\pi_\mu^{-1}\pi_R = \{(i\pi_R, j\pi_R) \mid i\pi_\mu > j\pi_\mu \text{ and } \mathrm{row}_{T_\mu}(i) > \mathrm{row}_{T_\mu}(j)\}$$

$$= \Gamma_1 \cup \Gamma_2 \cup \Gamma_3,$$

where $\Gamma_1 = \{(i\pi_R, j\pi_R) \mid i\pi_\mu > j\pi_\mu,\ i\pi_R \le r < j\pi_R, \text{ and } \mathrm{row}_{T_\mu}(i) > \mathrm{row}_{T_\mu}(j)\}$

$\Gamma_2 = \{(i\pi_R, j\pi_R) \mid i\pi_\mu > j\pi_\mu,\ r < j\pi_R < i\pi_R, \text{ and } \mathrm{row}_{T_\mu}(i) > \mathrm{row}_{T_\mu}(j)\}$

$\Gamma_3 = \{(i\pi_R, j\pi_R) \mid i\pi_\mu > j\pi_\mu,\ j\pi_R < i\pi_R \le r, \text{ and } \mathrm{row}_{T_\mu}(i) > \mathrm{row}_{T_\mu}(j)\}$,

by Lemma 12.2. But $\Gamma_1, \Gamma_2, \Gamma_3$ are disjoint closed subsets of Φ, so an application of Theorem 9.2 shows that

$$(12.4) \quad \bar{P}_\mu \pi_R \prod_{i=1}^{3} \Big(\sum_{g \in G(\Gamma_i)} \chi_{\pi_R^{-1}\pi_\mu c}(g)g \Big) \in S_\mu.$$

Next, recall that in Section 10 we defined $\sigma \sim \tau$ (for $\sigma, \tau \in \mathfrak{S}_n$) if $\mathrm{row}_{T_\mu \sigma}(i) = \mathrm{row}_{T_\mu \tau}(i)$ for all i with $1 \le i \le r$. For each $\sim$-class, $[\pi] = \{\pi' \mid \pi' \sim \pi\}$, let

$$M_{[\pi]} = \sum_{\substack{\pi' \sim \pi \\ u \in U^-}} \bar{K}(\bar{P}_\mu \pi' u).$$

Now, U^-G_{n-r} is a group (it is the set of matrices described in (10.3)), and by Lemma 10.7

(12.5) For each $\pi \in \mathfrak{S}_n$, $M_{[\pi]}$ is U^-G_{n-r}-invariant.

Given $\sigma, \tau \in \mathfrak{S}_n$, we write $[\sigma] < [\tau]$ if for some i_1, with $1 \le i_1 \le r$,

$\text{row}_{T_\mu\sigma}(i_1) < \text{row}_{T_\mu\tau}(i_1)$, but

$\text{row}_{T_\mu\sigma}(i) = \text{row}_{T_\mu\tau}(i)$ for all i with $i_1 < i \le r$.

We now aim to prove:

(12.6) If $1 \ne g \in G(\Gamma_1)$, then $P_\mu\pi_R g = P_\mu\tau u$ for some $\tau \in \mathfrak{S}_n$, $u \in U^-$, with $[\tau] < [\pi_R]$.

A non-singular matrix m belongs to P_μ if and only if $m_{ij} = 0$ when $\text{row}_{T_\mu}(i) > \text{row}_{T_\mu}(j)$. Therefore, given $\sigma, \tau \in \mathfrak{S}_n$,

$$(12.7)\quad m \in \sigma^{-1}P_\mu\tau \Leftrightarrow (\sigma m\tau^{-1})_{ij} = 0 \text{ when } \text{row}_{T_\mu}(i) > \text{row}_{T_\mu}(j)$$

$$\Leftrightarrow m_{ij} = 0 \text{ when } \text{row}_{T_\mu\sigma}(i) > \text{row}_{T_\mu\tau}(j).$$

Suppose that $1 \ne g \in G(\Gamma_1)$ and $P_\mu\pi_R g = P_\mu\tau u$ ($\tau \in \mathfrak{S}_n$, $u \in U^-$). Then $gu^{-1} \in \pi_R^{-1}P_\mu\tau$, and since $g \ne 1$, there exists (i, j) with $i \ne j$ and $g_{ij} \ne 0$. Among all such (i, j) choose (i_1, j_1) such that i_1 is maximal, and then j_1 is maximal. Since $g \in G(\Gamma_1)$,

$$(12.8)\quad i_1 \le r < j, \text{ and } \text{row}_{T_\mu\pi_R}(i_1) > \text{row}_{T_\mu\pi_R}(j_1).$$

The choice of (i_1, j_1) ensures that $(gu^{-1})_{i_1j_1} \ne 0$ and $(gu^{-1})_{ii} = 1$ for $i > i_1$. Bearing in mind that $gu^{-1} \in \pi_R^{-1}P_\mu\tau$, these results yield, by an application of (12.7) that

$$(12.9)\quad \text{row}_{T_\mu\pi_R}(i_1) \le \text{row}_{T_\mu\tau}(j_1)$$

$$\text{row}_{T_\mu\pi_R}(i) \le \text{row}_{T_\mu\tau}(i) \text{ for } i > i_1.$$

Let $\varepsilon = \mathrm{row}_{T_\mu\pi_R}(i_1)$. Since $i_1 \le r$, $\mathrm{row}_{T_\mu\pi_R}(b) < \varepsilon$ for all b with $b < i_1$, by the definition of π_R. Thus, if $\mathrm{row}_{T_\mu\pi_R}(b) \ge \varepsilon$, then $b \ge i_1$ and $b \ne j_1$ (using (12.8)). Conditions (12.9) now imply that

$$\{b \mid \mathrm{row}_{T_\mu\tau}(b) = \delta\} = \{b \mid \mathrm{row}_{T_\mu\pi_R}(b) = \delta\} \text{ if } \delta > \varepsilon,$$

and $$\{b \mid \mathrm{row}_{T_\mu\tau}(b) = \varepsilon\} = \{b \mid \mathrm{row}_{T_\mu\pi_R}(b) = \varepsilon\} \cup \{j_1\} \setminus \{i_1\}.$$

In particular,

$$\mathrm{row}_{T_\mu\tau}(i_1) < \mathrm{row}_{T_\mu\pi_R}(i_1)$$

and $$\mathrm{row}_{T_\mu\tau}(b) = \mathrm{row}_{T_\mu\pi_R}(b) \text{ for } i_1 < b \le r.$$

Therefore, $[\tau] < [\pi_R]$, and this completes the proof of (12.6).

Since $G(\Gamma_2)$, $G(\Gamma_3) \subseteq U^-$, when we apply (12.6) to the element of S_μ described in (12.4), and use (12.5), we obtain:

$$(12.10) \quad \bar{P}_\mu\pi_R \Big(\sum_{g\in G(\Gamma_2)} \chi_{\pi_R^{-1}\pi_\mu c}(g)g\Big)\Big(\sum_{g\in G(\Gamma_3)} \chi_{\pi_R^{-1}\pi_\mu c}(g)g\Big)$$

$$+ \text{(an element of } \sum_{[\pi]<[\pi_R]} M_{[\pi]}) \text{ belongs to } S_\mu.$$

But Γ_3 is a subset of $\Gamma(r)$ (see Definition 9.4), and for $(i, j) \in \Gamma_3$,

$$\begin{aligned}(i, j)\pi_R^{-1}\pi_\mu c &= 2 \quad \text{if } i = j + 1 \\ &= 1 \quad \text{if } i \ne j + 1,\end{aligned}$$

by Lemma 12.2(iii). Thus, for all $g \in G(\Gamma_3)$,

$$\chi_{\pi_R^{-1}\pi_\mu c}(g)gE_r = E_r.$$

Therefore, multiplying the element in (12.10) on the right by $q^{-|\Gamma_3|}E_r$, we obtain:

$$(12.11) \quad \bar{P}_\mu\pi_R \Big(\sum_{g\in G(\Gamma_2)} \chi_{\pi_R^{-1}\pi_\mu c}(g)g\Big)E_r$$

$$+ (\text{an element of } \sum_{[\pi]<[\pi_R]} M_{[\pi]}) \text{ belongs to } S_\mu E_r .$$

Under the isomorphisms of Theorems 10.12 and 10.14,

$$\bar{P}_\mu \pi_R E_r (\sum_{g \in G(\Gamma_2)} \chi_{\pi_R^{-1} \pi_\mu c} (g) g)$$

maps to a generator of S_{μ_R}, by condition (ii) in the definition 12.1 of T_R. Therefore, multiplying (12.11) on the right by appropriate elements of $\bar{K}G_{n-r}$ we can obtain ($\dim S_{\mu_R}$) linearly independent elements in $S_\mu E_r$ of the form

$$(\text{Element of } M_{[\pi_R]}) + (\text{element of } \sum_{[\pi]<[\pi_R]} M_{[\pi]})$$

(Here, we are using (12.5). Note that $G(\Gamma_2) \subseteq G_{n-r}$ and the elements of G_{n-r} commute with E_r).

As a vector space, $M_\mu = \bigoplus M_{[\pi]}$, the sum being over distinct $\sim$-classe (see Lemma 7.3), so we have proved that

$$\dim (S_\mu E_r) \geq \sum_{R \in \mathcal{R}_r^*} \dim S_{\mu_R} .$$ ■

Theorem 9.11(ii) now gives:

12.12 COROLLARY (cf. Theorem 10.21 and 16.9)

$$\dim S_\mu \geq \sum_{r=1}^{h} \sum_{R \in \mathcal{R}_r^*} (q^{n-1} - 1)(q^{n-2} - 1) \ldots (q^{n-r+1} - 1) \dim S_{\mu_R} .$$ ■

In fact, it is possible to extract more information about $S_\mu E_r$ from the proof of Theorem 12.3. To do this, for $R, R' \in \mathcal{R}_r^*$, define $R < R'$ if $[\pi_R] < [\pi_{R'}]$. That is, the set $\mathcal{R}_r^*$ is ordered by "last differences"; in other words,

(12.13) For distinct elements R, R' of $\mathcal{R}_r^*$, we write $R < R'$ if the largest element of $R \setminus R'$ is less than the largest element of $R' \setminus R$.

Now suppose that $\mathcal{R}_r^* = \{R_1 < R_2 < \ldots < R_k\}$ (thus $k = \binom{h}{r}$ if μ has distinct parts.) Let s_j be the element of $S_\mu E_r$ described in (12.11), taking $R = R_j$, and define

$$S_j = \sum_{i=1}^{j} s_i \bar{K}G_{n-r}.$$

Then the part of the proof of Theorem 12.3 after (12.11) proves

12.14 THEOREM (cf. Theorem 16.9)

$$S_\mu E_R \supseteq S_k \supset S_{k-1} \supset \cdots \supset S_1 \supset S_0 = 0,$$

<u>and for each j with $1 \le j \le k$, the exists a $\bar{K}G_{n-r}$-homomorphism θ_j such that</u>

$$S_j\theta_j = S_{\mu_{R_j}} \text{ and } S_{j-1}\theta_j = 0.$$

■

Our linearly independent elements of S_μ (which turn out to be a basis) use, via the construction of T_R and induction on n, μ-tableaux which tend to have the property that the numbers decrease along each row and increase down each column. One might expect to get μ-tableaux resembling standard ones by redefining G^*_{n-1} to be a subgroup of the stabilizer of the subspace $\langle e_n \rangle$, instead of $\langle e_1 \rangle$. However, closer examination reveals that this alternative does not work smoothly, and it is likely that some reversal of the numbers is intrinsic to the result.

Since dim S_μ (as we shall see) is a polynomial in q, where the sum of the coefficients equals the number of standard μ-tableaux, it is interesting to see combinatorially that a standard μ-tableau may be associated in a natural way with each power of q in the polynomial for dim S_μ (Thomas [T]). Can this association be used in constructing a basis of S_μ? Sometimes, but not always. For example, dim $S_{(2,2)} = q^2 + q^4$, and there is a natural (but difficult) way of constructing a basis where q^2 and q^4 elements are associated with the tableaux

$$\begin{matrix} 1 & 3 \\ 2 & 4 \end{matrix} \qquad \begin{matrix} 1 & 2 \\ 3 & 4 \end{matrix},$$

respectively. Similar methods applied to $S_{(3,3)}$ give, however,

$$q^3 \qquad q^5 \qquad q^5 \qquad q^7 + q^6 - q^5 \qquad q^9$$

basis elements associated, respectively, with the tableaux

$$\begin{matrix} 1 & 3 & 5 \\ 2 & 4 & 6 \end{matrix} \qquad \begin{matrix} 1 & 3 & 4 \\ 2 & 5 & 6 \end{matrix} \qquad \begin{matrix} 1 & 2 & 5 \\ 3 & 4 & 6 \end{matrix} \qquad \begin{matrix} 1 & 2 & 4 \\ 3 & 5 & 6 \end{matrix} \qquad \begin{matrix} 1 & 2 & 3 \\ 4 & 5 & 6 \end{matrix}.$$

This gives some indication of the difficulties involved in devising a basis for S_μ. In our opinion, Theorem 12.3 is the hardest result in the essay; this is in marked contrast to the situation for $\mathfrak{S}_n$, where a lower bound for the dimension of a Specht module is much easier to find than an upper bound.

13 THE KERNEL INTERSECTION THEOREM FOR $S_{(n-m,m)}$

We shall prove that S_μ equals the intersection of the kernels of certain $\bar{K}G_n$-homomorphisms $\psi_{d,i}$, defined on M_μ (when μ has h non-zero parts, the subscripts d, i are bounded by $1 \le d \le h-1$, $0 \le i \le \mu_{d+1} - 1$). The special case where $\mu = (n-m, m)$ with $n - m \ge m$ will be presented first, because there is an easy trick which works here; in general, the maps $\psi_{d,i}$ involve only adjacent parts of μ, so at least the full Kernel Intersection Theorem will look more plausible after this section! Another good reason for dealing first with $\mu = (n-m, m)$ is that the result for this case is sufficient to show in the next section that the parts of an arbitrary μ can be rearranged without loss.

Assume, throughout this section, that $n - m \ge m > 0$. From Corollary 12.12, we know that

$$(13.1)\quad \dim S_{(n-m,m)} \ge \dim S_{(n-m,m-1)} + \dim S_{(n-m-1,m)} + (q^{n-1}-1)\dim S_{(n-m-1,m-1)},$$

where the second term must be omitted if $n = 2m$.

We shall prove that we have equality here, and that

$$\dim S_{(n-m,m)} = \begin{bmatrix} n \\ m \end{bmatrix} - \begin{bmatrix} n \\ m-1 \end{bmatrix}.$$

By induction on n, we may assume that the number on the right-hand side of (13.1) equals

$$\left(\begin{bmatrix} n-1 \\ m-1 \end{bmatrix} - \begin{bmatrix} n-1 \\ m-2 \end{bmatrix}\right) + \left(\begin{bmatrix} n-1 \\ m \end{bmatrix} - \begin{bmatrix} n-1 \\ m-1 \end{bmatrix}\right) + (q^{n-1}-1)\left(\begin{bmatrix} n-2 \\ m-1 \end{bmatrix} - \begin{bmatrix} n-2 \\ m-2 \end{bmatrix}\right).$$

The second term is automatically zero if $n = 2m$. Rearranging, and applying Theorem 3.4, this number equals

$$\begin{bmatrix} n-1 \\ m-1 \end{bmatrix} + \begin{bmatrix} n-1 \\ m \end{bmatrix} + (q^m-1)\begin{bmatrix} n-1 \\ m \end{bmatrix} - \begin{bmatrix} n-1 \\ m-2 \end{bmatrix} - \begin{bmatrix} n-1 \\ m-1 \end{bmatrix} - (q^m-1)\begin{bmatrix} n-2 \\ m-1 \end{bmatrix}$$

$$= \begin{bmatrix} n \\ m \end{bmatrix} - \begin{bmatrix} n \\ m-1 \end{bmatrix}.$$

Therefore,

$$(13.2)\quad \dim S_{(n-m,m)} \geq \begin{bmatrix} n \\ m \end{bmatrix} - \begin{bmatrix} n \\ m-1 \end{bmatrix}.$$

We regard $M_{(n-m,m)}$ as the vector space over $\bar{K}$ which has a basis consis of all the m-dimensional subspaces of V, and define the KG_n-homomorphism $\psi_{1,}$ from $M_{(n-m,m)}$ into $M_{(n-i,i)}$ as in 2.16; that is, for $0 \leq i \leq m$, $\psi_{1,i}$ sends an m-dimensional space to the sum of all the i-dimensional spaces contained in it. Note that $\psi_{1,m}$ is the identity map.

13.3 THEOREM

$$S_{(n-m,m)} = \bigcap_{i=0}^{m-1} \operatorname{Ker} \psi_{1,i} \quad \underline{\text{and}} \quad \dim S_{(n-m,m)} = \begin{bmatrix} n \\ m \end{bmatrix} - \begin{bmatrix} n \\ m-1 \end{bmatrix}.$$

<u>Furthermore, if we define</u> $M_s = \bigcap_{i=0}^{s-1} \operatorname{Ker} \psi_{1,i}$ $(0 \leqq s \leqq m+1)$, <u>then</u>

$$M_{(n-m,m)} = M_0 \supset M_1 \supset M_2 \supset \ldots M_m \supset M_{m+1} = 0,$$

<u>and</u> $M_s/M_{s+1} \cong S_{(n-s,s)}$ $(0 \leq s \leq m)$.

<u>Proof</u>: Consider the generator ξ of $S_{(n-s,s)}$ $(0 \leq s \leq m)$, described in Examp 11.17(v). Let V_1 be the m-dimensional subspace of V spanned by

$$e_2, e_4, \ldots, e_{2s}, e_{2s+1}, \ldots, e_{m+s}.$$

The image of V_1 under $\psi_{1,s}$ is the sum of all the s-dimensional subspaces of V_1. But of these s-dimensional subspaces, precisely one, namely that spanne by

$$e_2, e_4, \ldots, e_{2s},$$

occurs in the expression for ξ. Therefore, $\langle V_1\psi_{1,s}, \xi\rangle = 1$, and so

$$M_0\psi_{1,s} \nsubseteq S^{\perp}_{(n-s,s)}.$$

By the Submodule Theorem, (11.12(ii)),

(13.4) $M_0\psi_{1,s} \supseteq S_{(n-s,s)}$, for $0 \le s \le m$.

At this stage, to avoid repeating the remainder of the proof later, we weaken our hypothesis as follows:

(13.5) Assume that M_0 is a submodule of $M_{(n-m,m)}$ such that

$$M_0\psi_{1,s} \supseteq S_{(n-s,s)}, \text{ for } 0 \le s \le m.$$

Redefine M_s for $0 \le s \le m+1$ by

$$M_s = M_0 \cap \left(\bigcap_{i=0}^{s-1} \operatorname{Ker} \psi_{1,i}\right).$$

We shall deduce from this hypothesis that $M_0 = M_{(n-m,m)}$, and the results in the statement of the theorem.

The next step is to prove:

(13.6) If $0 \le t \le s \le m$, then no composition factor of M_0/M_t is isomorphic to $D_{(n-s,s)}$.

This is certainly true if $t = 0$, so we assume that $t > 0$ and that $D_{(n-s,s)}$ is not a composition factor of M_0/M_{t-1}. But $M_0/(M_0 \cap \operatorname{Ker} \psi_{1,t-1})$ is isomorphic to a $\bar{K}G_n$-submodule of $M_{(n-t+1,t-1)}$, so it does not have $D_{(n-s,s)}$ as a composition factor, by Lemma 11.13. Since $M_t = M_{t-1} \cap \operatorname{Ker} \psi_{t-1}$, $D_{(n-s,s)}$ is not a composition factor of M_0/M_t. This proves (13.6).

From (13.6), we know that no composition factor of $(M_0\psi_{1,s})/(M_s\psi_{1,s})$ is isomorphic to $D_{(n-s,s)}$, for $0 \le s \le m$. But

$$\begin{aligned}(M_0\psi_{1,s})/(M_s\psi_{1,s}) &\supseteq (S_{(n-s,s)} + M_s\psi_{1,s})/(M_s\psi_{1,s}) \\ &\cong S_{(n-s,s)}/(S_{(n-s,s)} \cap M_s\psi_{1,s}).\end{aligned}$$

Therefore, $M_s\psi_{1,s} \not\supseteq S^{\perp}_{(n-s,s)}$, since

$$D_{(n-s,s)} = S_{(n-s,s)}/(S_{(n-s,s)} \cap S^{\perp}_{(n-s,s)}),$$

by definition. Thus,

$$M_s\psi_{1,s} \supseteq S_{(n-s,s)} \qquad \text{for } 0 \leq s \leq m,$$

by the Submodule Theorem. Furthermore, $M_m \supseteq S_{(n-m,m)}$, by Corollary 11.14(ii), so we have now proved the existence of submodules M_i' $(0 \leq i \leq m)$ of M_0 such that

$$M_0 \supseteq M_0' \supset M_1 \supseteq M_1' \supset M_2 \supseteq \ldots \supset M_m \supseteq M_m' \supset M_{m+1} = 0,$$

and $\quad M_s'/M_{s+1} \cong S_{(n-s,s)} \qquad \text{for } 0 \leq s \leq m.$

But

$$\begin{bmatrix} n \\ m \end{bmatrix} = \sum_{s=0}^{m} \left(\begin{bmatrix} n \\ s \end{bmatrix} - \begin{bmatrix} n \\ s-1 \end{bmatrix} \right) \leq \sum_{s=0}^{m} \dim S_{(n-s,s)}, \text{ by (13.2)},$$

$$= \sum_{s=0}^{m} \dim (M_s'/M_{s+1}) \leq \sum_{s=0}^{m} \dim(M_s/M_{s+1})$$

$$= \dim M_0 \leq \dim M_{(n-m,m)} = \begin{bmatrix} n \\ m \end{bmatrix}.$$

Therefore, we have equality in all possible places. In particular,

$$M_{(n-m,m)} = M_0,$$

$$M_s' = M_s \qquad \text{for } 0 \leq s \leq m,$$

and $\quad \dim S_{(n-s,s)} = \begin{bmatrix} n \\ s \end{bmatrix} - \begin{bmatrix} n \\ s-1 \end{bmatrix} \quad \text{for } 0 \leq s \leq m.$

This concludes the proof of the Theorem. ■

13.7 COROLLARY (see also Theorem 16.12)

<u>The trivial $\bar{K}G_n$-module is a submodule of $S_{(n-m,m)}$ if and only if the characteristic of K (= p_K, say) divides</u>

$$\begin{bmatrix} n \\ m \end{bmatrix}, \begin{bmatrix} n-1 \\ m-1 \end{bmatrix}, \ldots, \begin{bmatrix} n-m+2 \\ 2 \end{bmatrix}, \text{ and } \begin{bmatrix} n-m+1 \\ 1 \end{bmatrix}.$$

<u>Proof</u>: Clearly, the only trivial $\bar{K}G_n$-submodule of $M_{(n-m,m)}$ is spanned by the sum of all the m-dimensional subspaces of V. But this element of $M_{(n-m,}$

belongs to Ker $\psi_{1,i}$ if and only if for each i-dimensional subspace V_1 of V, p_K divides the number of m-dimensional subspace V_2 of V such that $V_2 \supset V_1$. That is, by Corollary 3.3(i),

$$p_K \text{ divides } \begin{bmatrix} n - i \\ m - i \end{bmatrix}.$$

Since $S_{(n-m,m)} = \bigcap_{i=0}^{m-1} \text{Ker } \psi_{1,i}$, the result follows. ■

We can now prove that when μ is a composition of n obtained by rearranging the parts of λ, then $M_\mu \cong M_\lambda$. To demonstrate that this is not trivial, we begin with an example to show that M_μ need not be isomorphic to M_λ if we drop the restriction that the characteristic of K does not divide q.

EXAMPLE. Let $n = 3$, $q = 2$, $K = \mathbb{F}_2$, $\lambda = (2, 1)$, $\mu = (1, 2)$. Let θ_0, θ_1 be the KG_n-homomorphisms from $M_{(1,2)}$ to $M_{(2,1)}$ such that θ_0 sends a 2-dimensional subspace of V to the sum of all the 1-dimensional subspaces of V, and θ_1 sends a 2-dimensional subspace to the sum of all the 1-dimensional subspaces which intersect it in zero. By considering orbits, it is clear that the only non-zero KG_n-homomorphisms from $M_{(1,2)}$ into $M_{(2,1)}$ are θ_0, θ_1 and $\theta_0 + \theta_1$.

Rank $\theta_0 = 1$, so θ_0 is not an isomorphism.

The matrix of θ_1 (with respect to the natural bases) is:

(14.1)

	$\langle e_1\rangle$	$\langle e_2\rangle$	$\langle e_3\rangle$	$\langle e_1+e_2\rangle$	$\langle e_1+e_3\rangle$	$\langle e_2+e_3\rangle$	$\langle e_1+e_2$
$\langle e_1, e_2\rangle$	0	0	1	0	1	1	1
$\langle e_1, e_3\rangle$	0	1	0	1	0	1	1
$\langle e_2, e_3\rangle$	1	0	0	1	1	0	1
$\langle e_1+e_2, e_3\rangle$	1	1	0	0	1	1	0
$\langle e_1+e_3, e_2\rangle$	1	0	1	1	0	1	0
$\langle e_2+e_3, e_1\rangle$	0	1	1	1	1	0	0
$\langle e_1+e_2, e_2+e_3\rangle$	1	1	1	0	0	0	1

Calculation reveals that the rank of this matrix (over $\mathbb{F}_2$) is 3, so rank $\theta_1 = 3$, and hence rank $(\theta_0 + \theta_1) \leq 4$. Therefore, neither θ_1 nor $\theta_0 + \theta_1$ is an isomorphism, and $M_{(1,2)} \not\cong M_{(2,1)}$.

Incidentally, suppose we were to define $S_{(2,1)}$, in this case where the characteristic of K divides q, to be the subset of $M_{(2,1)}$ consisting of those vectors for which the sum of the coefficients equals zero (cf. Theorem 13.3). Then $\mathrm{Im}\,\theta_1 \not\supseteq S_{(2,1)}$ and $\mathrm{Im}\,\theta_1 \not\subseteq S^{\perp}_{(2,1)}$, since $\dim \mathrm{Im}\,\theta_1 = 3$, $\dim S_{(2,1)} = 6$, $\dim S^{\perp}_{(2,1)} = 1$. Thus, the Submodule Theorem fails when the characteristic of K divides q. ■

We now adopt once more the hypothesis that the characteristic of K does not divide q, and continue to regard $M_{(n-s,s)}$ as the vector space over K spanned by the s-dimensional subspaces of V. Our first aim is to prove that $M_{(n-m,m)} \cong M_{(m,n-m)}$, and a preliminary Lemma is needed:

14.2 LEMMA. <u>Assume that $s \le \min(m, n-m)$, and let V_1 be an $(n-m)$-dimensional subspace of V. Then</u>

$$\Sigma\{V_2 \mid \dim V_2 = s \text{ and } V_2 \cap V_1 = 0\} \notin S^{\perp}_{(n-s,s)}.$$

<u>Proof</u>: By applying an appropriate element of G_n, we may assume that $V_1 = \langle e_2, e_4, \ldots, e_{2s}, e_{2s+1}, e_{2s+2}, \ldots, e_{n-m+s}\rangle$. Let ξ be the generator of $S_{(n-s,s)}$ described in Example 11.17(v). Thus, ξ is a linear combination of s-dimensional subspaces of $\langle e_1, e_2, \ldots, e_{2s}\rangle$. Of the terms appearing in ξ, precisely those where each $\varepsilon_{ij} = 1$ have zero intersection with $\langle e_2, e_4, \ldots, e_{2s}\rangle$, and hence have zero intersection with V_1. Therefore,

$$\langle \xi, \Sigma\{V_2 \mid \dim V_2 = s \text{ and } V_2 \cap V_1 = 0\}\rangle_{(n-s,s)}$$

$$= (-1)^s q^{s(s-1)/2} \neq 0.$$ ■

4.3 THEOREM. <u>$M_{(n-m,m)}$ and $M_{(m,n-m)}$ are isomorphic $\bar{K}G_n$-modules.</u>

<u>Proof:</u> We may assume that $m \le n - m$.

Let θ be the $\bar{K}G_n$-homomorphism from $M_{(m,n-m)}$ to $M_{(n-m,m)}$ which maps each $(n-m)$-dimensional subspace of V to the sum of the m-dimensional subspaces which intersect it in zero. We shall prove that θ is an isomorphism.

Assume that V_1 is an $(n-m)$-dimensional subspace of V. For $0 \le s \le m$, let $\psi_{1,s}$ be the $\bar{K}G_n$-homomorphism from $M_{(n-m,m)}$ to $M_{(n-s,s)}$ defined in 2.16. We claim that

(14.4) $V_1\theta\psi_{1,s} = q^{(n-m)(m-s)} \Sigma\{V_2 \mid \dim V_2 = s \text{ and } V_2 \cap V_1 = 0\}.$

Let V_2 be an s-dimensional subspace of V. If $V_2 \cap V_1 = 0$, then the number of m-dimensional subspaces V_3 of V such that $V_3 \supseteq V_2$ and $V_3 \cap V_1 = 0$ is $q^{(n-m)(m-s)}$ by Theorem 3.1. On the other hand, if $V_2 \cap V_1 \neq 0$, there is no subspace V_3 of V such that $V_3 \supseteq V_2$ and $V_3 \cap V_1 = 0$. This proves (14.4).

By Lemma 14.2, $V_1\theta\psi_{1,s} \notin S^{\perp}_{(n-s,s)}$. Therefore, applying the Submodule Theorem, we deduce that

(14.5) $M_{(m,n-m)}\theta\psi_{1,s} \supseteq S_{(n-s,s)}$ for $0 \leq s \leq m$.

Now, this shows that if we let $M_0 = M_{(m,n-m)}\theta$, then M_0 satisfies Hypothesis 13.5, from which we concluded that $M_0 = M_{(n-m,m)}$. But $\dim M_{(m,n-m)} = \dim M_{(n-m,m)}$, so θ is an isomorphism, as we wished to show. ■

We have proved that the map θ which sends each (n - m)-dimensional subspace V_1 of V to the element of $M_{(n-m,m)}$ consisting the sum of the m-dimensional subspaces which intersect V_1 in zero is an isomorphism when $2m \leq n$. This shows that the elementary divisors of the matrix 14.1 (viewed as an integer-valued matrix) are all powers of 2; they are, in fact, 1, 1, 1, 2, 2, 2, 4. It seems remarkable that every matrix constructed in a similar way must have elementary divisors which are powers of the prime which divides q.

Since θ is an isomorphism,

(14.6) For each m-dimensional subspace V* of V (with $2m \leq n$), there is a $\bar{K}$ linear combination of (n - m)-dimensional subspaces of V, say ξ, such that $\xi\theta = V^*$.

14.7 THEOREM. <u>If μ is a composition of n obtained from λ by permuting its</u>

parts, then M_λ and M_μ are isomorphic $\bar{K}G_n$-modules.

Proof: It is sufficient to prove the theorem in the case where μ is obtained by interchanging two adjacent parts of λ, so we assume that

$$\lambda = (\lambda_1, \lambda_2, \ldots, \lambda_d, \lambda_{d+1}, \ldots, \lambda_h)$$

$$\mu = (\lambda_1, \lambda_2, \ldots, \lambda_{d+1}, \lambda_d, \ldots, \lambda_h) \text{ and } \lambda_d \le \lambda_{d+1}$$

The elements of M_λ may be regarded as $\bar{K}$ linear combinations of λ-flags

$$V = V_0 \supseteq V_1 \supseteq \cdots \supseteq V_{d-1} \supseteq V_d \supseteq V_{d+1} \supseteq \cdots \supseteq V_h = 0$$

Define θ to be the $\bar{K}G_n$-homomorphism which sends the above λ-flag to the sum of all the μ-flags which have the form

$$V = V_0 \supseteq V_1 \supseteq \cdots \supseteq V_{d-1} \supseteq W_d \supseteq V_{d+1} \supseteq \cdots \supseteq V_h = 0,$$

with

$$W_d \cap V_d = V_{d+1}.$$

In view of (14.6), for each μ-flag like

$$V = V_0 \supseteq V_1 \supseteq \cdots \supseteq V_{d-1} \supseteq W_d^* \supseteq V_{d+1} \supseteq \cdots \supseteq V_h = 0,$$

there is a $\bar{K}$ linear combination of λ-flags of the form

$$V = V_0 \supseteq V_1 \supseteq \cdots \supseteq V_{d-1} \supseteq V_d' \supseteq V_{d+1} \supseteq \cdots \supseteq V_h = 0$$

which is sent by θ to this particular μ-flag. Thus, θ maps M_λ onto M_μ, and since $\dim M_\lambda = \dim M_\mu$, we have proved that $M_\lambda \cong M_\mu$. ∎

It now looks very likely that $S_\lambda \cong S_\mu$ (and hence also $D_\lambda \cong D_\mu$); if we knew that $S_\lambda\theta \nsubseteq S_\mu^\perp$, the result would come out using the Submodule Theorem, or if we knew that D_μ is not a composition factor of M_λ/S_λ, we could again deduce what we want. However, neither of these results is straightforward yet, so we postpone the proof that $S_\lambda \cong S_\mu$ to Section 16, when it is easy.

15 THE KERNEL INTERSECTION THEOREM

We have seen already that S_λ is contained in the intersection of the kernels of all $\bar{K}G_n$-homomorphisms which map M_λ into some M_μ with $\mu \triangleright \lambda$. In this section we shall prove equality, and thereby bring together several results from earlier chapters. Actually, we shall prove that S_λ equals the intersection of the kernels of several carefully selected $\bar{K}G_n$-homomorphisms.

15.1 DEFINITION. Suppose that $\lambda = (\lambda_1, \lambda_2, \ldots, \lambda_h)$, and assume that d < and $0 \leq i \leq \lambda_d$. Define

$$\mu = (\lambda_1, \lambda_2, \ldots, \lambda_{d-1}, \lambda_d + \lambda_{d+1} - i, i, \lambda_{d+2}, \ldots, \lambda_h),$$

so that $\mu_j = \lambda_j$ if $j \neq d, d+1$. Choose right coset representatives $g_1, g_2,$ for $P_\lambda \cap P_\mu$ in P_λ. Let $\psi_{d,i}$ be the $\bar{K}G_n$-homomorphism from M_λ into M_μ defined by

$$\psi_{d,i}: \quad \bar{P}_\lambda \xi \mapsto \bar{P}_\mu (\sum_j g_j) \xi \qquad (\xi \in \bar{K}G_n).$$

It is easy to describe $\psi_{d,i}$ in terms of λ-flags; it sends the flag

$$V = V_0 \supsetneq V_1 \supsetneq \cdots \supsetneq V_{d-1} \supsetneq V_d \supsetneq V_{d+1} \supsetneq \cdots \supsetneq V_h = 0$$

to the sum of all the flags

$$V = V_0 \supsetneq V_1 \supsetneq \cdots \supsetneq V_{d-1} \supsetneq W_d \supsetneq V_{d+1} \supsetneq \cdots \supsetneq V_h = 0$$

which have the property that

$$V_d \supseteq W_d \quad \text{and} \quad \dim (W_d/V_{d+1}) = i .$$

In particular, the notation here is consistent with Definition 2.16. We proved in Theorem 13.3 that when $\lambda = (\lambda_1, \lambda_2)$ with $\lambda_1 \geq \lambda_2$, we have

$$S_\lambda = \bigcap_{i=0}^{\lambda_2 - 1} \operatorname{Ker} \psi_{1,i}.$$

The general Kernel Intersection Theorem deals with arbitrary compositions but when $\lambda = (\lambda_1, \lambda_2, \ldots, \lambda_h)$ is a partition, we shall have

$$S_\lambda = \bigcap_{d=2}^{h} \bigcap_{i=0}^{\lambda_d - 1} \operatorname{Ker} \psi_{d-1,i}.$$

Besides proving that S_λ is a kernel intersection, we shall also describe generators for many other kernel intersections which are contained in M_λ (not, we may add, through any desire for generality, but because we have found no other route to the result for S_λ.) We are guided here by our own work on the symmetric groups (cf. James [J_4]).

For a start, we define "a pair of compositions $\lambda^\#$, λ for n" and a corresponding $\bar{K}G_n$-submodule $S_{\lambda^\#,\lambda}$ of M_λ such that $S_{\lambda^\#,\lambda} \supseteq S_\lambda$.

15.2 DEFINITION. Let $\lambda = (\lambda_1, \lambda_2, \ldots, \lambda_h)$ be a composition of n and let $\lambda^\# = (\lambda_1^\#, \lambda_2^\#, \ldots, \lambda_h^\#)$ be a <u>partition</u> of some $n' \le n$, such that

$\lambda_i^\# \le \lambda_i$ for all i

$0 < \lambda_i^\# < \lambda_i$ for at most one i.

We shall call $\lambda^\#$, λ a <u>pair of compositions for n</u>. If $\lambda^\# \ne \lambda$, we let $d = d(\lambda^\#)$ be the least non-negative integer such that $\lambda_d^\# < \lambda_d$.

In examples, we shall denote a pair of compositions for n by drawing the diagram for λ with lines between the rows, then boxing in the part corresponding to $\lambda^\#$.

EXAMPLE. If $\lambda = (3, 2, 3, 2)$ and $\lambda^\# = (3, 2, 1, 0)$, then we draw:

X X X
X X
X X X
X X

If T is a λ-tableau, then $T^\#$ denotes the part of T which is boxed in during this process. Thus, in the example above,

if T =
3 8 4
6 7
2 1 9
5 10
then $T^\#$ =
3 8 4
6 7
2 .

15.3 DEFINITION. Given a pair of compositions $\lambda^{\#}$, λ for n (with $d = d(\lambda^{\#})$) let

$$\Gamma_{\lambda^{\#},\lambda} = \{(i, j) \mid i, j \in (T_\lambda\pi_\lambda)^{\#} \text{ and } i > j\}$$

$$\cup \{(i, j) \mid \text{row}_{T_\lambda\pi_\lambda}(i) = \text{row}_{T_\lambda\pi_\lambda}(j) = d, \text{ and}$$

$$\lambda_d^{\#} < \text{col}_{T_\lambda\pi_\lambda}(j) < \text{col}_{T_\lambda\pi_\lambda}(i)\}$$

$$\cup \{(i, j) \mid \text{row}_{T_\lambda\pi_\lambda}(i) = d, \ j \in (T_\lambda\pi_\lambda)^{\#}, \text{ and}$$

$$\text{col}_{T_\lambda\pi_\lambda}(j) \leqq \lambda_d^{\#} < \text{col}_{T_\lambda\pi_\lambda}(i)\}.$$

The definition of $T_\lambda\pi_\lambda$ ensures that $\Gamma_{\lambda^{\#},\lambda} \subseteq \Phi^-$.

Let us explain why we take this complicated definition of $\Gamma_{\lambda^{\#},\lambda}$. The first set is natural enough; it equals Φ^- when $\lambda^{\#} = \lambda$. In the proof of (15.8) below, we need

$$\{(i, j) \mid i > j \text{ and } \text{row}_{T_\lambda\pi_\lambda}(i) = \text{row}_{T_\lambda\pi_\lambda}(j) = d\}$$

to be a subset of $\Gamma_{\lambda^{\#},\lambda}$, and once we are committed including this set, we have to take $\Gamma_{\lambda^{\#},\lambda}$ in the form given, in order to ensure that $\Gamma_{\lambda^{\#},\lambda}$ is a closed subset of Φ^-.

EXAMPLE. If $\lambda = (3, 2, 4)$ and $\lambda^{\#} = (3, 2, 1)$ then

$$T_\lambda\pi_\lambda = \begin{array}{|c|}\hline 1\ 4\ 7 \\ \hline 2\ 5 \\ \hline 3 \,|\, 6\ 8\ 9 \\ \hline\end{array},$$

and $\Gamma_{\lambda^{\#},\lambda} = \{(i, j) \mid i, j \in \{1\ 2\ 3\ 4\ 5\ 7\} \text{ and } i > j\}$

$$\cup \{(8\ 6), (9\ 6), (9\ 8)\}$$

$$\cup \{(i, j) \mid i \in \{6\ 8\ 9\}, \ j \in \{1\ 2\ 3\}\}.$$

15.4 DEFINITION (cf. 9.1 and 11.4).

Let c be the function from $\Gamma_{\lambda^{\#},\lambda}$ to $\{1, 2, \ldots, q\}$ defined by

$(i, j)c = 2$, if $i = j + 1$ and $\mathrm{col}_{T_\lambda \pi_\lambda}(i) = \mathrm{col}_{T_\lambda \pi_\lambda}(j)$

$(i, j)c = 1$, otherwise .

Let

$$E_{\lambda^\#,\lambda} = \frac{1}{q^{|\Gamma_{\lambda^\#,\lambda}|}} \sum_{g \in G(\Gamma_{\lambda^\#,\lambda})} \chi_c(g) g.$$

15.5 DEFINITION. Let $S_{\lambda^\#,\lambda} = \bar{P}_\lambda \pi_\lambda E_{\lambda^\#,\lambda} (\bar{K}G_n)$.

If $\lambda^\# = (0, 0, \ldots, 0)$, then $\Gamma_{\lambda^\#,\lambda}$ is empty, $E_{\lambda^\#,\lambda} = 1$, and so $S_{\lambda^\#,\lambda} = M_\lambda$.

If $\lambda^\# = \lambda$, then λ is a partition, and the definition of $\Gamma_{\lambda^\#,\lambda}$ ensures that $S_{\lambda^\#,\lambda} = S_\lambda$.

By Theorem 10.2,

$\bar{P}_\lambda \pi_\lambda E_{\lambda^\#,\lambda}$ is a non-zero multiple of $\bar{P}_\lambda \pi_\lambda \sum_{g \in G(\Gamma)} \chi_c(g) g$, where

$$\Gamma = \{(i, j) \mid i, j \in (T_\lambda \pi_\lambda)^\#,\ i > j, \text{ and } \mathrm{row}_{T_\lambda \pi_\lambda}(i) > \mathrm{row}_{T_\lambda \pi_\lambda}(j)\}$$

$$\cup \{(i, j) \mid \mathrm{row}_{T_\lambda \pi_\lambda}(i) = d > \mathrm{row}_{T_\lambda \pi_\lambda}(j),\ j \in (T_\lambda \pi_\lambda)^\#, \text{ and}$$
$$\mathrm{col}_{T_\lambda \pi_\lambda}(j) \leq \lambda_d^\# < \mathrm{col}_{T_\lambda \pi_\lambda}(i)\}$$

$$= \{(i, j) \mid i, j \in (T_\lambda \pi_\lambda)^\#,\ \mathrm{row}_{T_\lambda \pi_\lambda}(i) > \mathrm{row}_{T_\lambda \pi_\lambda}(j), \text{ and}$$
$$\mathrm{col}_{T_\lambda \pi_\lambda}(j) \leq \mathrm{col}_{T_\lambda \pi_\lambda}(i)\}$$

$$\cup \{(i, j) \mid \mathrm{row}_{T_\lambda \pi_\lambda}(i) = d > \mathrm{row}_{T_\lambda \pi_\lambda}(j) \text{ and}$$
$$\mathrm{col}_{T_\lambda \pi_\lambda}(j) \leq \lambda_d^\# < \mathrm{col}_{T_\lambda \pi_\lambda}(i)\}.$$

Now assume that $d = d(\lambda^\#) = 1$, and consider the above set. If we define $\nu^\# = (\nu_1^\#, \nu_2^\#, \ldots, \nu_h^\#)$, by

$\nu_1^\# = \lambda_1 \qquad \nu_i^\# = \lambda_i^\#$ for $i > 1$, then $\nu^{\#}_{d'} = 0$, where $d' = d(\nu^{\#})$, since $0 < \lambda^{\#}_i < \lambda_i$ for at most one i. Hence the set Γ above is the same as the set we get when $\lambda^{\#},\lambda$ is replaced by $\nu^{\#},\lambda$. Thus $S_{\lambda^{\#},\lambda} = S_{\nu^{\#},\lambda}$ when $d(\lambda^{\#}) = 1$. This proves

(15.6) If $d(\lambda^{\#}) = 1$, then $S_{\lambda^{\#},\lambda} = S_{\nu^{\#},\lambda}$, where $\nu^{\#}$ is obtained from $\lambda^{\#}$ by absorbing the first part of λ.

Therefore, when we are investigating $S_{\lambda^{\#},\lambda}$, we may assume that $d(\lambda^{\#}) > 1$; in particular, there is little loss in writing this hypothesis into the next two theorems. The conclusion of each of these theorems concerns an inclusion between two $\bar{K}G_n$-modules; afterwards we prove, by very indirect means, that we have equality.

15.7 THEOREM. <u>Assume that $\lambda^{\#}$, λ is a pair of compositions for n, with $\lambda^{\#} \neq \lambda$ and $d = d(\lambda^{\#}) > 1$. Let $\mu = (\mu_1, \mu_2, \ldots)$ be defined by</u>

$$\mu_{d-1} = \lambda_{d-1} + \lambda_d - \lambda^{\#}_d$$

$$\mu_d = \lambda^{\#}_d$$

$$\mu_i = \lambda_i \quad \text{for } i \neq d-1, d.$$

<u>Then</u>

$$S_{\lambda^{\#},\lambda}\,\psi_{d-1,\lambda^{\#}_d} \supseteq S_{\lambda^{\#},\mu}.$$

<u>Proof</u>: First of all, note that $\lambda^{\#}$, μ is a pair of compositions for n; it is obtai from $\lambda^{\#}$, λ by removing the nodes of row d which are not boxed in, and placing them back at the end of row (d - 1). Also, $M_\lambda \psi_{d-1,b} \subseteq M_\mu$, by definition, where for notational convenience we write $b = \lambda^{\#}_d$.

<u>Step 1</u> We prove that $\bar{P}_\lambda \pi_\lambda E_{\lambda^{\#},\lambda} \psi_{d-1,b} = \bar{P}_\mu \pi_\lambda E_{\lambda^{\#},\lambda}$.

Since $\mu_d < \lambda_d$, $P_\lambda \cap P_\mu = P_\nu$ where

$$\nu = (\lambda_1, \lambda_2, \ldots, \lambda_{d-1}, \lambda_d - \mu_d, \mu_d, \lambda_{d+1}, \ldots, \lambda_h).$$

The right coset representatives of $P_\lambda \cap P_\mu$ in P_λ, used in the definition of $\psi_{d-1,b}$, may therefore be taken in the form

$$\tau_k u_k \qquad (\tau_k \in W,\ u_k \in U^-,\ 1 \le k \le m, \text{ say})$$

where

$$\tau_1 u_1 = 1,$$

and for all i and k,

$$i\tau_k = i \quad \text{if } \mathrm{row}_{T_\lambda}(i) \ne d,$$

and $\quad u_k \in \langle X_{ij} \mid i > j \text{ and } \mathrm{row}_{T_\lambda}(i) = \mathrm{row}_{T_\lambda}(j) = d\rangle.$

Then

$$\bar{P}_\lambda \pi_\lambda \psi_{d-1,b} = \sum_{k=1}^{m} \bar{P}_\mu \tau_k \pi_\lambda (\pi_\lambda^{-1} u_k \pi_\lambda).$$

But

$$\pi_\lambda^{-1} u_k \pi_\lambda \in \langle X_{ij} \mid i\pi_\lambda^{-1} > j\pi_\lambda^{-1} \text{ and } \mathrm{row}_{T_\lambda\pi_\lambda}(i) = \mathrm{row}_{T_\lambda\pi_\lambda}(j) = d\rangle$$

$$= \langle X_{ij} \mid i > j \text{ and } \mathrm{row}_{T_\lambda\pi_\lambda}(i) = \mathrm{row}_{T_\lambda\pi_\lambda}(j) = d\rangle,$$

by the definition of π_λ.

So

(15.8) $\pi_\lambda^{-1} u_k \pi_\lambda \in G(\Gamma_{\lambda^\#,\lambda})$.

Thus, $\bar{P}_\lambda \tau_k u_k \pi_\lambda E_{\lambda^\#,\lambda}$ is a non-zero multiple of $\bar{P}_\mu \tau_k \pi_\lambda E_{\lambda^\#,\lambda}$.

Now, by Theorem 10.2,

$$\bar{P}_\mu \tau_k \pi_\lambda E_{\lambda^\#,\lambda} \ne 0$$

$$\Leftrightarrow \{(i, j) \mid i = j + 1,\ \mathrm{col}_{T_\lambda\pi_\lambda}(i) = \mathrm{col}_{T_\lambda\pi_\lambda}(j), \text{ and } i, j \in (T_\lambda\pi_\lambda)^\#\}$$

$$\subseteq \{(i, j) \mid \mathrm{row}_{T_\mu\tau_k\pi_\lambda}(i) > \mathrm{row}_{T_\mu\tau_k\pi_\lambda}(j)\}$$

$$\Leftrightarrow \ \{(i, j) \mid i, j \in T_\lambda^{\#}, \ row_{T_\lambda}(i) = row_{T_\lambda}(j) + 1, \text{ and } col_{T_\lambda}(i) = col_{T_\lambda}$$

$$\subseteq \{(i, j) \mid row_{T_\mu \tau_k}(i) > row_{T_\mu \tau_k}(j)\}$$

$$\Leftrightarrow \ \{(i, j) \mid row_{T_\lambda}(i) = d, \ row_{T_\lambda}(j) = d - 1, \text{ and } col_{T_\lambda}(i) = col_{T_\lambda}(j)$$

$$\subseteq \{(i, j) \mid row_{T_\mu \tau_k}(i) = d \text{ and } row_{T_\mu \tau_k}(j) = d - 1\},$$

since T_λ and T_μ agree on all rows except rows $(d - 1)$ and d.

But τ_k permutes only the numbers in row d of T_λ, so $\bar{P}_\mu \tau_k \pi_\lambda E_{\lambda^{\#},\lambda}$ is non-zero if and only if τ_k stabilizes row d of T_μ. However, in that case, $\tau_k \in P_\lambda \cap P_\mu$, and $k = 1$. We have now proved that

$$\bar{P}_\lambda \pi_\lambda E_{\lambda^{\#},\lambda} \psi_{d-1,b} = \bar{P}_\mu \pi_\lambda E_{\lambda^{\#},\lambda}.$$

<u>Step 2</u> Let T be the μ-tableau obtained from $T_\lambda \pi_\lambda$ by moving those numbers in row d which lie outside $(T_\lambda \pi_\lambda)^{\#}$ to the end of row $(d - 1)$.

e.g. if $T_\lambda \pi_\lambda = \begin{array}{l} 1\ 5\ 9 \\ 2\ 6 \\ 3\ 7\ 10 \\ 4\ 8 \end{array}$ then $T = \begin{array}{l} 1\ 5\ 9 \\ 2\ 6\ 7\ 10 \\ 3 \\ 4\ 8 \end{array}$

while $T_\mu \pi_\mu = \begin{array}{l} 1\ 5\ 8 \\ 2\ 6\ 9\ 10 \\ 3 \\ 4\ 7 \end{array}$.

We describe $\Gamma_{\lambda^{\#},\lambda}$ as a set of pairs of numbers in T:

$$\Gamma_{\lambda^{\#},\lambda} = \{(i, j) \mid i, j \in (T_\lambda \pi_\lambda)^{\#} \text{ and } i > j\}$$

$$\cup \{(i, j) \mid row_T(i) = row_T(j) = d - 1, \text{ and } \lambda^{\#}_{d-1} < col_T(j) < col_T(i)\}$$

$$\cup \{(i, j) \mid row_T(i) = d - 1, \ j \in (T_\lambda \pi_\lambda)^{\#}, \text{ and } col_T(j) \leq \lambda^{\#}_d \leq \lambda_{d-1} < col_T(i)\}.$$

Let $\sigma \in \mathfrak{S}_n$ be defined by $T\sigma = T_\mu \pi_\mu$. Then

$$\Gamma_{\lambda^{\#},\lambda}\sigma = \{(i, j) \mid i, j \in (T_\mu \pi_\mu)^{\#} \text{ and } i > j\}$$

$$\cup \{(i, j) \mid \mathrm{row}_{T_\mu \pi_\mu}(i) = \mathrm{row}_{T_\mu \pi_\mu}(j) = d - 1, \text{ and }$$

$$\lambda^{\#}_{d-1} < \mathrm{col}_{T_\mu \pi_\mu}(j) < \mathrm{col}_{T_\mu \pi_\mu}(i)\}$$

$$\cup \{(i, j) \mid \mathrm{row}_{T_\mu \pi_\mu}(i) = d - 1, \ j \in (T_\mu \pi_\mu)^{\#}, \text{ and }$$

$$\mathrm{col}_{T_\mu \pi_\mu}(j) \le \lambda^{\#}_d \le \lambda_{d-1} < \mathrm{col}_{T_\mu \pi_\mu}(i)\}.$$

Comparing this with the definition of $\Gamma_{\lambda^{\#},\mu}$, we see that

$$\Gamma_{\lambda^{\#},\lambda}\sigma \subseteq \Gamma_{\lambda^{\#},\mu}.$$

(In general, the third subset of $\Gamma_{\lambda^{\#},\lambda}\sigma$ may be strictly contained in the third set in the definition of $\Gamma_{\lambda^{\#},\mu}$.)

Therefore, by Theorem 9.2,

$$\sigma^{-1} E_{\lambda^{\#},\lambda} \sigma\xi = E_{\lambda^{\#},\mu} \qquad \text{for some } \xi \in \bar{K}G_n.$$

Then, applying the result of Step 1, we have:

$$\bar{P}_\lambda \pi_\lambda E_{\lambda^{\#},\lambda} \sigma\xi\psi_{d-1,b} = \bar{P}_\mu \pi_\lambda \sigma E_{\lambda^{\#},\mu}.$$

Finally, it is easy to see that $T_\mu \pi_\mu$ differs from $T_\mu \pi_\lambda \sigma$ only by a row permutation, so $\bar{P}_\mu \pi_\lambda \sigma = \bar{P}_\mu \pi_\mu$. Thus, we have constructed an element of $S_{\lambda^{\#},\lambda}$ whose image under $\psi_{d-1,b}$ is a generator of $S_{\lambda^{\#},\mu}$, and the theorem follows. ■

It is worth noting that the symmetric group result which corresponds to this one (see James [J_6], Theorem 17.13) has equality in the conclusion. Fortunately, the inclusion in Theorem 15.7 is the way round which enables us later to prove that it is an equality.

15.9 THEOREM. <u>Assume that $\lambda^{\#}$, λ is a pair of compositions for n, with $\lambda^{\#} \neq \lambda$ and $d = d(\lambda^{\#}) > 1$, and that $\lambda^{\#}_d < \lambda^{\#}_{d-1}$. Let $\mu^{\#} = (\mu^{\#}_1, \mu^{\#}_2, \ldots)$ be defined by</u>

$$\mu_d^{\#} = \lambda_d^{\#} + 1$$

$$\mu_i^{\#} = \lambda_i^{\#} \quad \text{for } i \neq d.$$

<u>Then</u>

$$S_{\lambda^{\#},\lambda} \cap \operatorname{Ker} \psi_{d-1,\lambda_d^{\#}} \supseteq S_{\mu^{\#},\lambda}.$$

<u>Proof</u>: First of all, note that $\mu^{\#}, \lambda$ is a pair of compositions for n, since $\lambda_d^{\#} < \lambda_{d-1}^{\#}$.

Now, $\Gamma_{\lambda^{\#},\lambda} \subseteq \Gamma_{\mu^{\#},\lambda}$, by construction, so

$$E_{\lambda^{\#},\lambda}\, \xi = E_{\mu^{\#},\lambda} \quad \text{for some } \xi \in \bar{K}G_n,$$

and this proves that $S_{\lambda^{\#},\lambda} \supseteq S_{\mu^{\#},\lambda}$.

$\psi_{d-1,b}$ maps M_λ into M_μ $(b = \lambda_d^{\#})$, where μ is the composition defined in Theorem 15.7, and as in Step 1 of the proof of that theorem, $S_{\mu^{\#},\lambda}\,\psi_{d-1,b}$ is non-zero only if, for some k,

$$\bar{P}_\mu \tau_k \pi_\lambda E_{\mu^{\#},\lambda} \neq 0 .$$

But for this we need

$$\{(i, j) \mid \text{row}_{T_\lambda}(i) = d,\ \text{row}_{T_\lambda}(j) = d - 1, \text{ and } \text{col}_{T_\lambda}(i) = \text{col}_{T_\lambda}(j) \le \lambda_d^{\#} + 1\}$$

$$\subseteq \{(i, j) \mid \text{row}_{T_\mu \tau_k}(i) = d \text{ and } \text{row}_{T_\mu \tau_k}(j) = d - 1\},$$

and this is impossible (the first set has cardinality $\lambda_d^{\#} + 1$ and the second has cardinality $\lambda_d^{\#}$). Therefore, $S_{\mu^{\#},\lambda}\,\psi_{d-1,b} = 0$. ■

The last two theorems are illustrated by:

EXAMPLE. If $\lambda^{\#}, \lambda =$

```
X X X ,
X X X
X X X X
X
```

then $\psi_{1,1}$ maps $S_{\lambda^{\#},\lambda}$ onto a module containing $S_{\lambda^{\#},\mu}$, where

$$\lambda^{\#},\ \mu = \begin{array}{l} \boxed{X\ X\ X}\ X\ X \\ \boxed{X} \\ X\ X\ X\ X \\ X \end{array}$$

and $S_{\lambda^{\#},\lambda} \cap \text{Ker}\ \psi_{1,1}$ contains $S_{\mu^{\#},\lambda}$, where

$$\mu^{\#},\ \lambda = \begin{array}{l} \boxed{X\ X\ X} \\ \boxed{X\ X}\ X \\ X\ X\ X\ X \\ X \end{array} \ .$$

By (15.6), $S_{\lambda^{\#},\mu} = S_{\nu^{\#},\mu}$, where

$$\nu^{\#},\ \mu = \begin{array}{l} \boxed{X\ X\ X\ X\ X} \\ \boxed{X} \\ X\ X\ X\ X \\ X \end{array} .$$

Now, transforming $\lambda^{\#}$, λ to $\lambda^{\#}$, μ always moves some nodes up, and when they reach row 1 they can be boxed in. On the other hand, transforming $\lambda^{\#}$, λ to $\mu^{\#}$, λ boxes in an extra node. Repeated applications of such manoeuvres to $\lambda^{\#}$, λ always lead to a situation where all the nodes are boxed in.

To formalise this process, we give:

15.10 DEFINITION. Suppose that $\lambda^{\#}$, λ is a pair of compositions for n, with $\lambda^{\#} \neq \lambda$. We define two new pairs of compositions $(\lambda^{\#}, \lambda)I$ and $(\lambda^{\#}, \lambda)K$ for n as follows (I stands for "Image", K for "Kernel"):

(i) If $d(\lambda^{\#}) = 1$, let $(\lambda^{\#}, \lambda)I = \nu^{\#}, \lambda$, where

$$\nu_1^{\#} = \lambda_1 \qquad \nu_i^{\#} = \lambda_i \quad \text{for } i > 1 \text{ (cf. 15.6)}.$$

If $d(\lambda^{\#}) > 1$, let $(\lambda^{\#}, \lambda)I = \lambda^{\#}, \mu$, as defined in Theorem 15.7.

(ii) We define $(\lambda^{\#}, \lambda)K$ only if $d(\lambda^{\#}) > 1$ and $\lambda_d^{\#} < \lambda_{d-1}^{\#}$. In this case, let $(\lambda^{\#}, \lambda)K = \mu^{\#}, \lambda$, as defined in Theorem 15.9.

EXAMPLE

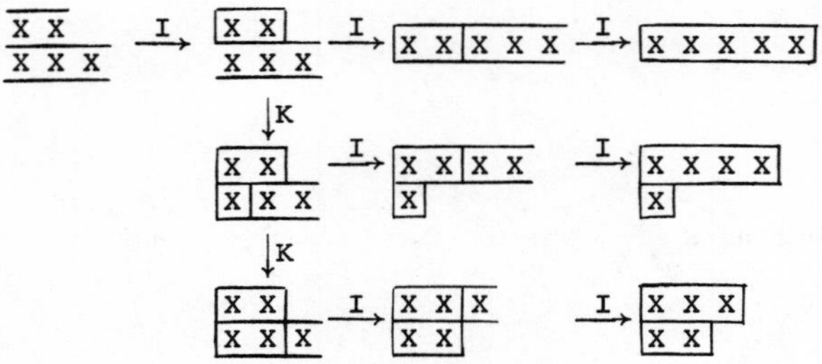

```
 X X         I    [X X]        I    [X X|X X X]   I    [X X X X X]
 X X X      -->    X X X      -->                -->
                   |K
                   v
                  [X X]        I    [X X|X X       I    [X X X X]
                  [X|X X       -->  [X]           -->  [X]
                   |K
                   v
                  [X X]        I    [X X|X         I    [X X X]
                  [X X|X]     -->   [X X]         -->  [X X]
```

The only pairs of compositions to which we can apply neither I nor K are those where $\lambda^{\#} = \lambda$ (and so λ is a partition of n). For every other pair of compositions, we can apply the maps I and K successively to reach a pair of compositions $\mu^{\#}$, μ, where $\mu^{\#} = \mu$. The example above shows that from the pair $\lambda^{\#} = (0, 0)$, $\lambda = (2, 3)$ we can get to μ, μ, where μ is one of the partitions (5), (4, 1), (3, 2). In order to describe the partitions we reach (together with multiplicities), we give:

15.11 DEFINITION. If λ is a composition of n and μ is a partition of n, then a semistandard μ-tableau of type λ is an array of integers obtained by replacing the nodes in the diagram $[\mu]$ by integers (allowing repeats) such that

(i) The numbers are strictly increasing down each column.

(ii) The numbers are non-decreasing along each row.

(iii) For each $i \geq 1$, the number of entries equal to i is λ_i.

EXAMPLE

```
1 1 2 2 3
2 2 3
3 3
```

is a semistandard (5, 3, 2)-tableau, of type (2, 4, 4).

15.12 LEMMA. *Let $\lambda^{\#}$, λ be a pair of compositions for n, and μ be a partition of n. Then the number of distinct sequences*

$$J_1 \; J_2 \; \ldots \qquad \text{(each } J_k = I \text{ or } K\text{)}$$

such that

$$(\lambda^{\#}, \lambda) J_1 J_2 \ldots = \mu, \mu$$

equals the number of semistandard μ-tableau T of type λ which have the property that for each i, every number in row i of $T^{\#}$ is i.

Proof: The result is certainly true when $\lambda^{\#} = \lambda$, so assume that $\lambda^{\#} \neq \lambda$ and that the result is true for $(\lambda^{\#}, \lambda)I$ and also for $(\lambda^{\#}, \lambda)K$, if it is defined.

The lemma is then trivial if $d(\lambda^{\#}) = 1$; in this case, all sequences $J_1 \; J_2 \ldots$ start with I, and a semistandard tableau of type λ has λ_1 1's in row 1. Assume, therefore, that $d = d(\lambda^{\#}) > 1$.

The statement of the lemma may be rephrased by saying that the μ, μ which arise by applying sequences of I's and K's to $\lambda^{\#}$, λ are the partitions we get by first replacing the nodes in row i of $\lambda^{\#}$ by i, for all i, then adding $(\lambda_j - \lambda_j^{\#})$ numbers j, for all j, to create a semistandard tableau. In this process, all the $(\lambda_d - \lambda_d^{\#})$ numbers d which are to be added must go in row d or higher, by the definition of "semistandard". If we put one of these d's in row d we end up with those semistandard tableaux which arise from $(\lambda^{\#}, \lambda)K$. If we do not put one of these d's in row d, then all of them must go in row (d - 1) or higher. Since all the (d - 1)'s are already in row (d - 1) (because $\lambda_{d-1} = \lambda_{d-1}^{\#}$), we get the same shaped tableaux by renaming the d's, which are to be added, as d - 1. Thus, we get the tableaux which arise from $(\lambda^{\#}, \lambda)I$. Hence the result.

■

EXAMPLE. If $\lambda^{\#}, \lambda =$

X X X ,
X X
X X

the various tableaux described in the Lemma, together with the sequences leading to them are:

1 1 1 2 3 3
2
from I^5

1 1 1 2 3
2 3
from I^3KI^2

1 1 1 2
2 3 3
from I^3K^2

1 1 1 2 3
2
3
from I^2KI^3

1 1 1 2
2 3
3
from I^2KIK

1 1 1 3 3
2 2
from KI^3

1 1 1 3
2 2 3
from $KIKI^2$

1 1 1 3
2 2
3
from K^2I^3

1 1 1
2 2 3
3
from K^2IK

1 1 1
2 2
3 3
from K^3.

15.13 DEFINITION. If λ is a composition of n and μ is a partition of n, then $K_{\mu\lambda}$ is defined to be the number of semistandard μ-tableaux of type λ ($K_{\mu\lambda}$ is a Kostka number).

For example, if we write 0 for the unique composition of zero, then Lemma 15.12 shows that the number of distinct sequences $J_1\ J_2 \ldots$ (each $J_k = I$ or K) such that

$$(0,\ \lambda) J_1\ J_2 \ldots = \mu,\ \mu$$

equals $K_{\mu\lambda}$.

15.14 LEMMA. (cf. Theorems 13.3 and 15.16)

There is a series of $\bar{K}G_n$-submodules of M_λ,

$$M_\lambda = M_0 \supseteq M_0' \supset M_1 \supseteq M_1' \supset \ldots \supseteq M_t' \supset M_{t+1} \supseteq 0,$$

such that each M_s'/M_{s+1} is isomorphic to some S_μ, where μ is a partition of n. The number of factors M_s'/M_{s+1} which are isomorphic to S_μ equals $K_{\mu\lambda}$.

Proof: This follows from repeated applications of Theorems 15.7 and 15.9, and the remark preceding the statement of the lemma. ∎

The following well-known (and rather startling) fact now enables us to tie everything together:

(15.15) Let $x_1, x_2, \ldots$ be commuting indeterminates, and define $x_0 = 1$, $x_i = 0$ for $i < 0$. For each partition $\mu = (\mu_1, \mu_2, \ldots, \mu_h)$ of n, define

$$y_\mu = \det\,(x_{\mu_i + j - i})$$

(the determinant of an $h \times h$ matrix). Then for each composition λ of n,

$$x_{\lambda_1} x_{\lambda_2} \ldots = \sum_\mu K_{\mu\lambda} y_\mu .$$

EXAMPLE. If $\lambda = (2, 3)$, then the semistandard tableaux of type λ are

```
1 1 2 2 2     1 1 2 2     1 1 2 .
              2           2 2
```

Hence

$$x_2 x_3 = y_{(5)} + y_{(4,1)} + y_{(3,2)}$$

$$= |x_5| + \begin{vmatrix} x_4 & x_5 \\ x_0 & x_1 \end{vmatrix} + \begin{vmatrix} x_3 & x_4 \\ x_1 & x_2 \end{vmatrix} .$$

15.16 THEOREM

(i) *If μ is a partition of n, then (cf. Definition 10.17)*

$$\dim S_\mu = \{n\} \det \left(\frac{1}{\{\mu_i + j + i\}} \right) .$$

(ii) *If λ is a composition of n, then there is a series of $\bar{K}G_n$-submodules of M_λ,*

$$M_\lambda = M_0 \supset M_1 \supset M_2 \supset \ldots \supset M_t \supset M_{t+1} = 0$$

such that for each s $(0 \leq s \leq t)$, M_s/M_{s+1} is isomorphic to some S_μ where μ is a partition of n. The number of factors M_s/M_{s+1} which are isomorphic to S_μ equals $K_{\mu\lambda}$.

Proof: Note that the claimed dimension of S_μ is the number dim (μ), given in Definition 10.19. We assume, inductively, that dim S_ν = dim (ν) when ν is a composition of $n' < n$ (this is certainly true when $n' = 0$).

By Lemma 15.14,

$$\dim M_\lambda \geqq \sum_\mu K_{\mu\lambda} \dim S_\mu,$$

the sum being over all partitions μ of n. Then, applying Corollary 12.12, Theorem 10.21, and our induction hypothesis, we get:

$$\dim M_\lambda \geqq \sum_\mu K_{\mu\lambda} \dim S_\mu \geqq \sum_\mu K_{\mu\lambda} \dim (\mu).$$

But, if we put $x_i = \{i\}^{-1}$ in (15.15), and multiply the last equation therein by $\{n\}$, we obtain

$$\dim M_\lambda = \sum_\mu K_{\mu\lambda} \dim (\mu).$$

Therefore all the inequalities earlier in the proof are equalities and the theorem follows. ■

15.17 COROLLARY. Adopting the notation of Theorems 15.7 and 15.9, we have:

(i) $S_{\lambda^\#,\lambda}\psi_{d-1,\lambda_d^\#} = S_{\lambda^\#,\mu}$,

and this is an isomorphism if $\lambda_d^\# = \lambda_{d-1}^\#$.

(ii) $S_{\lambda^\#,\lambda} \cap \operatorname{Ker} \psi_{d-1,\lambda_d^\#} = S_{\mu^\#,\lambda}$ if $\lambda_d^\# < \lambda_{d-1}^\#$.

Proof: First note that any zero parts in a composition of n may be ignored.

It is easy to see that there exists a composition ν and a sequence $J_1 J_2 \ldots$ (each J_k = I or K) such that $(0, \nu)J_1 J_2 \ldots = (\lambda^\#, \lambda)$. The Corollary now follows by applying Theorem 15.16(ii) to M_ν. ■

EXAMPLE If $\lambda^{\#}, \lambda = $

```
X X X X
X X X|X X
X
X X
```

, then we take

$\nu = $

```
X X X X ,
X X X
X
X X
X X
```

whereupon $(0, \nu)IK^4I^2 = \lambda^{\#}, \lambda$. We deduce that M_ν has a subquotient isomorphic to $S_{\lambda^{\#},\lambda}$. ■

Another immediate corollary of Theorem 15.16 (since $S_{\mu^{\#},\mu} = S_\mu$ when $\mu^{\#} = \mu$) is:

15.18 THE KERNEL INTERSECTION THEOREM FOR PARTITIONS

<u>If $\mu = (\mu_1, \mu_2, \ldots, \mu_h)$ is a partition of n, then</u>

$$S_\mu = \bigcap_{d=2}^{h} \bigcap_{i=0}^{\mu_d - 1} \operatorname{Ker} \psi_{d-1,i}.$$

Combining this with Corollary 11.14(iii), we have certainly proved the next result in the case where λ is a partition. A little thought reveals that it has been proved for arbitrary compositions λ, but at the beginning of the next section we shall given an alternative proof.

15.19 THE KERNEL INTERSECTION THEOREM

<u>If λ is any composition of n, then</u>

$$S_\lambda = \bigcap_{\theta} \operatorname{Ker} \theta,$$

<u>the intersection being over all $\bar{K}G_n$-homomorphisms which map M_λ into some M_μ with $\mu \rhd \lambda$.</u>

Several important consequences follow quickly from the results proved in the last chapter.

Recall first that if μ is the partition of n obtained from λ by rearranging the parts in non-increasing order, then M_μ and M_λ are isomorphic $\bar{K}G_n$-modules; this was proved in Chapter 14.

16.1 THEOREM. Let λ be a composition of n, and let μ be the partition of n obtained by rearranging the parts of λ. If θ is any $\bar{K}G_n$-isomorphism from M_μ to M_λ, then $S_\mu\theta = S_\lambda$. In particular, S_λ and S_μ are isomorphic $\bar{K}G_n$-modules, and D_λ and D_μ are isomorphic $\bar{K}G_n$-modules.

At the same time as proving this theorem, we shall show

16.2 THEOREM. Let λ be a composition of n. Then every composition factor of M_λ/S_λ, and every composition factor of $S_\lambda \cap S_\lambda^\perp$, is isomorphic to some D_ν with $\nu \rhd \lambda$.

Proof: Theorem 16.2 is certainly true when $\lambda = (n)$, for in that case M_λ/S_λ and $S_\lambda \cap S_\lambda^\perp$ are both zero. We may therefore assume that Theorem 16.2 is corre[ct] for all compositions ω such that $\omega \rhd \lambda$.

We prove Theorem 16.2 first for the partition μ which is obtained by rearranging the parts of λ.

All the homomorphisms involved in the Kernel Intersection Theorem for the partition μ map M_μ into some M_ω with $\omega \rhd \mu$, and, by induction, every composition factor of M_ω (being either D_ω or a factor of M_ω/S_ω or $S_\omega \cap S_\omega^\perp$) has the form D_ν with $\nu \unrhd \omega$. Therefore, every composition factor of M_μ/S_μ is isomorphic to some D_ν with $\nu \rhd \mu$, But $S_\mu^\perp$ is isomorphic to the dual of M_μ/S_μ, and since every D_ν is self-dual, every composition factor of $S_\mu^\perp$ has the form D_ν with $\nu \rhd \mu$. This completes the proof of Theorem 16.2 for the partition μ.

Now let θ be a $\bar{K}G_n$-isomorphism from M_μ to M_λ. Since no composition factor of $M_\mu\theta/S_\mu\theta$ is isomorphic to D_λ, by what we have just proved, while $(S_\lambda + S_\lambda^\perp)/S_\lambda^\perp$ is isomorphic to D_λ, it follows that $S_\mu\theta \nsubseteq S_\lambda^\perp$. Therefore, $S_\mu\theta \supseteq S_\lambda$, by the Submodule Theorem.

On the other hand, no composition factor of $S_\mu^\perp$ is isomorphic to D_λ, either. Therefore, $S_\lambda \nsubseteq S_\mu^\perp\theta$, so $S_\lambda \supseteq S_\mu\theta$, by the Submodule Theorem applied to $M_\mu\theta$.

We have now proved that $S_\mu\theta = S_\lambda$. Thus, every composition factor of M_λ/S_λ, and hence also of $S_\lambda^\perp$, must be isomorphic to D_ν for some $\nu \rhd \lambda$.

Finally, since D_λ is the unique top composition factor of S_λ, it must be $\bar{K}G_n$-isomorphic to D_μ. ■

Theorem 16.1 shows immediately that the Kernel Intersection Theorem 15.19 for an arbitrary composition λ follows from the special case where λ is a partition.

16.3 COROLLARY. <u>Let λ be a composition of n. Then D_λ is a composition factor of S_λ (and of M_λ) with composition multiplicity one, and all other composition factors have the form D_ν, where $\nu \rhd \lambda$.</u>

<u>Proof</u>: Consider the following picture, and apply Theorem 16.2:

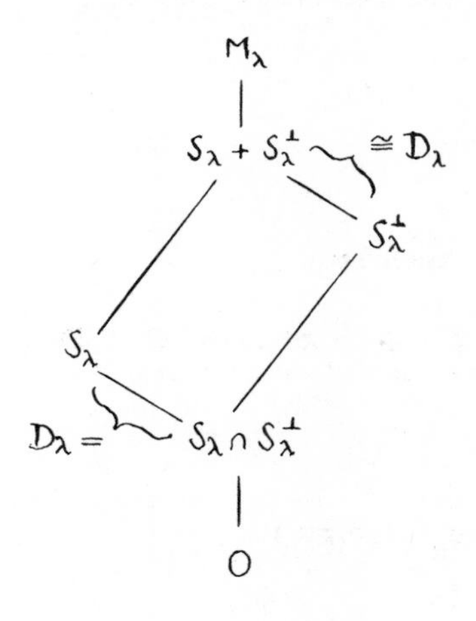

■

16.4 COROLLARY. Every irreducible unipotent representation of KG_n is isomorphic to some D_μ, where μ is a partition of n.

Proof: By definition, the irreducible unipotent modules are composition factors of the permutation representation of KG_n on B^+. But this permutation representation is $M_{(1,1,\ldots,1)}$, so the Corollary follows from Theorem 16.1 and Corollary 16.3.

■

16.5 COROLLARY. Dim S_λ is independent of the field K.

Proof: Combine Theorems 15.16(i) and 16.1.

■

16.6 THEOREM. If the prime p does not divide q, then the character of $S_{\lambda,\mathbb{Q}}$ (see Section 11), when restricted to p-regular classes of G_n, equals the p-modular character of $S_{\lambda,\mathbb{F}_p}$.

Proof: We may regard $\bar{P}_\lambda$ as an element of $\mathbb{Z}G_n$ ($\mathbb{Z}$ = the ring of integers) and each homomorphism $\psi_{d,i}$, defined in 15.1, sends $\bar{P}_\lambda\mathbb{Z}G_n$ into some $\bar{P}_\mu\mathbb{Z}G_n$ with $\mu \rhd \lambda$. The intersection of the kernels of several such maps may be viewed as a single kernel, by combining the maps into one $\mathbb{Z}G_n$-homomorphism

$$\psi^{\mathbb{Z}}: \quad \bar{P}_\lambda\mathbb{Z}G_n \to \bigoplus_{\mu\rhd\lambda} \bar{P}_\mu\mathbb{Z}G_n.$$

The Kernel Intersection Theorem shows that $S_{\lambda,\mathbb{Q}}$ equals $\mathbb{Q}\otimes \operatorname{Ker}\psi^{\mathbb{Z}}$ for some such map $\psi^{\mathbb{Z}}$. But

$$\psi^{\mathbb{Z}}: \quad p\bar{P}_\lambda\mathbb{Z}G_n \to p \bigoplus_{\mu\rhd\lambda} \bar{P}_\mu\mathbb{Z}G_n,$$

and hence we obtain a map $\psi^{\mathbb{F}_p}$, such that

$$\psi^{\mathbb{F}_p}: \quad (\bar{P}_\lambda\mathbb{Z}G_n)/(p\bar{P}_\lambda\mathbb{Z}G_n) \to (\bigoplus_{\mu\rhd\lambda} \bar{P}_\mu\mathbb{Z}G_n)/(p \bigoplus_{\mu\rhd\lambda} \bar{P}_\mu\mathbb{Z}G_n).$$

If $\alpha \in \bar{P}_\lambda\mathbb{Z}G_n$, write

$$\bar{\alpha} = \alpha + (p\bar{P}_\lambda\mathbb{Z}G_n) \in (\bar{P}_\lambda\mathbb{Z}G_n)/(p\bar{P}_\lambda\mathbb{Z}G_n).$$

If $\alpha \in \operatorname{Ker} \psi^{\mathbb{Z}}$, then $\bar{\alpha} \in \operatorname{Ker} \psi^{\mathbb{F}_p}$. Therefore,

$$\overline{\operatorname{Ker} \psi^{\mathbb{Z}}} \subseteq \operatorname{Ker} \psi^{\mathbb{F}_p}.$$

However, by Corollary 16.5

$$\dim (\operatorname{Ker} \psi^{\mathbb{F}_p}) = \dim (\mathbb{Q} \otimes \operatorname{Ker} \psi^{\mathbb{F}_p})$$

$$= \dim (\overline{\operatorname{Ker} \psi^{\mathbb{Z}}}).$$

Thus, $\overline{\operatorname{Ker} \psi^{\mathbb{Z}}} = \operatorname{Ker} \psi^{\mathbb{F}_p}$. Since $S_{\lambda,\mathbb{Q}} = \mathbb{Q} \otimes \operatorname{Ker} \psi^{\mathbb{Z}}$, and $S_{\lambda,\mathbb{F}_p} = \operatorname{Ker} \psi^{\mathbb{F}_p}$, the stated result follows.

■

Combined with Theorem 16.2, the last theorem proves:

16.7 THEOREM. <u>For primes p which do not divide q, part of the p-modular decomposition matrix of G_n has rows and columns indexed by partitions λ of n, and has the form</u>

$$\begin{array}{c} \\ (n) \\ (n-1,1) \\ \cdot \\ \cdot \\ \cdot \\ (1^n) \end{array} \begin{array}{c} \begin{array}{cccc} (n) & (n-1,1) & \cdots & (1^n) \end{array} \\ \left(\begin{array}{cccc} 1 & & & \\ & 1 & & 0 \\ & & \ddots & \\ & * & & \\ & & & 1 \end{array} \right) \end{array}$$

<u>The entry in row λ and column μ is zero unless $\mu \trianglerighteq \lambda$.</u>

Because we now know that, when μ is a partition of n, $\dim S_\mu = \dim (\mu)$, as in Definition 10.19, it follows from Theorem 12.3 and Corollary 12.12 that

(16.8) $$\dim (S_\mu E_r) = \sum_{R \in \mathcal{R}_r^*} \dim S_{\mu_R} \qquad \text{for } 1 \le r \le h$$

$$\dim S_\mu = \sum_{r=1}^{h} \sum_{R \in \mathcal{R}_r^*} (q^{n-1} - 1)(q^{n-2} - 1) \ldots (q^{n-r+1} - 1) \dim S_{\mu_R}.$$

From Theorem 12.14, we get

16.9 THEOREM. <u>Suppose that μ is a partition of n. Then there is a series</u>

of $\bar{K}G_{n-r}$-submodules of $S_\mu E_r$,

$$S_\mu E_r = S_k \supset S_{k-1} \supset \ldots \supset S_1 \supset S_0 = 0,$$

such that for each j with $1 \leq j \leq k$, S_j/S_{j-1} is $\bar{K}G_{n-r}$-isomorphic for $S_{\mu_{R_j}}$ (Here $R_1, \ldots, R_k$ are the elements of $\mathcal{R}^*_r$, ordered as in 12.13.)

In particular, by Theorem 9.11,

16.10 THE BRANCHING THEOREM.

If $\mu = (\mu_1, \mu_2, \ldots, \mu_h)$ is a partition of n, then as $\bar{K}G^*_{n-1}$-modules,

$$S_\mu = S_\mu E_1 \oplus \sum_{r=2}^{h} S_\mu E_r G^*_{n-1},$$

and $S_\mu E_1$ has a series of $\bar{K}G_{n-1}$-submodules,

$$S_\mu E_1 = S_h \supseteq S_{h-1} \supseteq \ldots \supseteq S_1 \supseteq S_0 = 0,$$

where

$$S_j = S_{j-1} \quad \text{if} \quad \mu_j = \mu_{j+1}$$

and S_j/S_{j-1} is $\bar{K}G_n$-isomorphic to

$$S_{(\mu_1, \mu_2, \ldots, \mu_j - 1, \ldots, \mu_h)} \quad \text{if} \quad \mu_j > \mu_{j+1}.$$

Compare the Branching Theorem with Corollary 10.16. Again, the analo with the symmetric group is clear; the term

$$\sum_{r=2}^{h} S_\mu E_r G^*_{n-1}$$

has dimension divisible by q - 1, so it "disappears when we put q = 1". For the symmetric group, the restriction of M_μ to $\mathfrak{S}_{n-1}$ is a direct sum of the form

$$\bigoplus_{R \in \mathcal{R}_1} M_{\lambda_R},$$

but the best we can do when restricting a Specht module to $\mathfrak{G}_{n-1}$ (except for some special fields, such as those of characteristic zero) is to construct a series of submodules like those in the description of $S_\mu E_1$ appearing in the Branching Theorem.

By considering the effect of E_1 on S_μ, we obtain from the Branching Theorem:

16.11 THEOREM. Let χ_λ denote the p-modular character of $S_{\lambda,\mathbb{Q}}$ (where p does not divide q). Suppose that the integers a_μ have the property that $\sum_\mu a_\mu \chi_\mu$ (summed over the partitions μ of n) is a p-modular character of G_n. Then $\sum_\mu a_\mu \left(\sum_{R \in \mathcal{R}_1^*} \chi_{\mu_R}\right)$ is zero or a p-modular character of G_{n-1}.

EXAMPLE. $\chi_{(1,1,1)} - \chi_{(2,1)}$ is not a p-modular character of G_3, if p is coprime to q, since $\chi_{(1,1)} - \chi_{(2)} - \chi_{(1,1)} = -\chi_{(1,1)}$ is not a p-modular character of G_2.

The Kernel Intersection Theorem is a very powerful tool when investigating the composition factors of S_μ, for it gives a straightforward test to determine whether or not a particular element of M_λ belongs to S_μ. We illustrate this by:

16.12 THEOREM (see also Corollary 13.7 and Theorem 19.6).

Suppose that $\mu = (\mu_1, \mu_2, \ldots, \mu_h)$ is a partition of n. The trivial $\bar{K}G_n$-module is a submodule of S_μ if and only if the characteristic of K divides

$$\begin{bmatrix} \mu_d + \mu_{d+1} - i \\ \mu_{d+1} - i \end{bmatrix}$$

for each d, i with $1 \le d \le h-1$, $0 \le i \le \mu_{d+1} - 1$.

Proof: Regard M_μ as the vector space over $\bar{K}$ spanned by μ-flags. Then the only trivial $\bar{K}G_n$-submodule of M_μ is that spanned by the sum of all the μ-fla[gs.] If $1 \le d \le h - 1$ and $0 \le i \le \mu_{d+1} - 1$, then the sum of all the μ-flags belon[gs] to Ker $\psi_{d,i}$ if and only if for each $(\mu_1, \mu_2, \ldots, \mu_{d-1}, \mu_d + \mu_{d+1} - i, i, \mu_{d+2}, \ldots, \mu_h)$-flag,

$$V = V_0 \supset V_1 \supset \ldots \supset V_{d-1} \supset U \supseteq V_{d+1} \supset \ldots \supset V_h = 0,$$

the characteristic of K divides the number of μ-flags of the form

$$V = V_0 \supset V_1 \supset \ldots \supset V_{d-1} \supset V_d \supset V_{d+1} \supset \ldots \supset V_h = 0$$

such that $V_d \supset U$. But the number of such flags is

$$\begin{bmatrix} \mu_d + \mu_{d+1} - i \\ \mu_{d+1} - i \end{bmatrix}$$

by Corollary 3.3(i). The result is now given by the Kernel Intersection Theorem 15.18. ■

Using the Kernel Intersection Theorem, we have also been able to pro[ve] the following theorem on hook partitions:

16.13 THEOREM. <u>Suppose that $1 \le m \le n - 1$, and let $\lambda = (n - m + 1, 1^{m-1})$, $\mu = (n - m, 1^m)$. If the characteristic of K divides [n], then there exists a non-zero $\bar{K}G_n$-homomorphism from S_λ into S_μ. Hence some composition factor of S_μ is isomorphic to D_λ.</u> ■

For the symmetric group, there is a theorem of Peel [P] which states that, when the characteristic of K divides n, there is a non-zero $K\mathfrak{S}_n$-homomorphism from the Specht module for λ into that for μ.

It can be shown that for $n \le 4$, each S_λ is irreducible unless the characteristic of K divides [1][2] ... [n]. Using Theorems 16.11, 12, 13, [and] further calculations, we have found that the part of the decomposition matr[ix] of G_n corresponding to unipotent representations is:

n = 2

$$\begin{array}{c} \\ (2) \\ (1,1) \end{array} \begin{array}{c} \begin{array}{cc} (2) & (1,1) \end{array} \\ \begin{pmatrix} 1 & 0 \\ 1 & 1 \end{pmatrix} \end{array}$$ if char K divides [2] = 1 + q

n = 3

$$\begin{array}{c|ccc} & (3) & (2,1) & (1,1,1) \\ \hline (3) & 1 & 0 & 0 \\ (2,1) & 0 & 1 & 0 \\ (1,1,1) & 1 & 0 & 1 \end{array}$$ if char K divides [2] = 1 + q

$$\begin{array}{c|ccc} & (3) & (2,1) & (1,1,1) \\ \hline (3) & 1 & 0 & 0 \\ (2,1) & 1 & 1 & 0 \\ (1,1,1) & 0 & 1 & 1 \end{array}$$ if char K divides $[3] = 1 + q + q^2$

n = 4

$$\begin{array}{c|ccccc} & (4) & (3,1) & (2,2) & (2,1^2) & (1^4) \\ \hline (4) & 1 & 0 & 0 & 0 & 0 \\ (3,1) & 1 & 1 & 0 & 0 & 0 \\ (2,2) & 0 & 1 & 1 & 0 & 0 \\ (2,1^2) & 1 & 1 & 1 & 1 & 0 \\ (1^4) & 1 & 0 & 1 & 1 & 1 \end{array}$$ if char K divides [2] = 1 + q.

$$\begin{array}{c|ccccc} & (4) & (3,1) & (2,2) & (2,1^2) & (1^4) \\ \hline (4) & 1 & 0 & 0 & 0 & 0 \\ (3,1) & 0 & 1 & 0 & 0 & 0 \\ (2,2) & 1 & 0 & 1 & 0 & 0 \\ (2,1^2) & 0 & 0 & 0 & 1 & 0 \\ (1^4) & 0 & 0 & 1 & 0 & 1 \end{array}$$ if char K divides $[3] = 1 + q + q^2$

$$\begin{array}{c|ccccc} & (4) & (3,1) & (2,2) & (2,1^2) & (1^4) \\ \hline (4) & 1 & 0 & 0 & 0 & 0 \\ (3,1) & 1 & 1 & 0 & 0 & 0 \\ (2,2) & 0 & 0 & 1 & 0 & 0 \\ (2,1^2) & 0 & 1 & 0 & 1 & 0 \\ (1^4) & 0 & 0 & 0 & 1 & 1 \end{array}$$

if char K divides $1 + q^2$, but not $1 + q$

In each matrix, the row labelled λ corresponds to S_λ and the column labelled μ corresponds to D_μ; thus the entry in row λ and column μ is the composition multiplicity of D_μ in S_λ. Compare these matrices with those in Example 1.7. The order of the partitions has been reversed, but this could be a matter of notation. Each of the above matrices describes the decomposition of ordinary representations of $GL_n(q)$ ($n \le 4$) over a field of characteristic coprime to q. Replacing [r] by r ("putting q = 1") we obtain matrices concerned with representations of $GL_d(F)$, for all d, over the field F of characteristic r. We conjecture that this is part of a general phenome - a really astonishing one, at that!

17 REMOVING THE FIRST COLUMN FROM $[\lambda]$

We are in a position to prove a theorem which relates the composition multiplicities of some of the modules D_μ in S_λ to composition multiplicities in smaller general linear groups.

The integer $h \le n$ will be fixed throughout this section. For every partition μ of n which has precisely h non-zero parts, we define the partition $\hat{\mu}$ of $n - h$ by

$$\hat{\mu} = (\mu_1 - 1, \mu_2 - 1, \ldots, \mu_h - 1).$$

Thus, the diagram for $\hat{\mu}$ is obtained from that for μ by removing the first column, and $M_{\hat{\mu}}$, $S_{\hat{\mu}}$, and $D_{\hat{\mu}}$ are $\bar{K}G_{n-h}$-modules. We shall prove:

17.1 THEOREM <u>If λ and μ are partitions of n, each having precisely h non-zero parts, then the composition multiplicity of D_μ in S_λ equals the composition multiplicity of $D_{\hat{\mu}}$ in $S_{\hat{\lambda}}$.</u>

The symmetric group result corresponding to this one has only recently been proved (James $[J_8]$), and it is interesting to note that our proof of the theorem for $\mathfrak{S}_n$ used Weyl modules; we know of no proof which lies entirely within the framework of the representation theory of symmetric groups.

Recall the Definition 9.4 of E_h. The elements of G_{n-h} commute with E_h, so ME_h is a $\bar{K}G_{n-h}$-module, for every $\bar{K}G_{n-h}$-module M. We have seen (Theorem 10.5) that

(17.2) If ν is a composition of n with fewer than h non-zero parts, then $M_\nu E_h = 0$. Hence, also, $S_\nu E_h = D_\nu E_h = 0$.

The main step in proving Theorem 17.1 is:

17.3 THEOREM. <u>If λ is a partition of n with precisely h non-zero parts then</u>

$$S_\lambda E_h \cong S_{\hat{\lambda}} \quad \underline{\text{and}} \quad D_\lambda E_h \cong D_{\hat{\lambda}},$$

the isomorphisms being $\bar{K}G_{n-h}$-isomorphisms.

Proof: By Theorem 15.16, there is a series of $\bar{K}G_n$-submodules of M_λ,

(17.4) $M_\lambda = M_0 \supset M_1 \supset \ldots \supset M_t \supset M_{t+1} = 0$,

such that $M_t = S_\lambda$ and for $0 \le s \le t-1$, $M_s/M_{s+1} \cong S_\mu$ ($\bar{K}G_n$-isomorphism) for some $\mu \rhd \lambda$. The number of factors M_s/M_{s+1} which are isomorphic to S_μ equals $K_{\mu\lambda}$.

In a similar way, there is a series of $\bar{K}G_{n-h}$-submodules of $M_{\hat{\lambda}}$,

(17.5) $M_{\hat{\lambda}} = \hat{M}_0 \supset \hat{M}_1 \supset \ldots \supset \hat{M}_u \supset \hat{M}_{u+1} = 0$,

such that $\hat{M}_u = S_{\hat{\lambda}}$ and for $0 \le s \le u-1$, $\hat{M}_s/\hat{M}_{s+1} \cong S_{\hat{\mu}}$ ($\bar{K}G_{n-h}$-isomorphism) for some $\hat{\mu} \rhd \hat{\lambda}$. The number of factors $\hat{M}_s/\hat{M}_{s+1}$ which are isomorphic to $S_{\hat{\mu}}$ equals $K_{\hat{\mu}\hat{\lambda}}$.

The definition of Kostka numbers (15.13) shows that $K_{\mu\lambda} = K_{\hat{\mu}\hat{\lambda}}$ if both λ and μ have precisely h non-zero parts. Thus,

(17.6) If μ has precisely h non-zero parts, then the number of factors M_s/M_{s+1} in (17.4) which are isomorphic to S_μ equals the number of factors $\hat{M}_s/\hat{M}_{s+1}$ in (17.5) which are isomorphic to $S_{\hat{\mu}}$.

We next assume, inductively, that if μ is a partition of n with precisely h non-zero parts and $\mu \rhd \lambda$, then $S_\mu E_h \cong S_{\hat{\mu}}$ and $D_\mu E_h \cong D_{\hat{\mu}}$ as $\bar{K}G_{n-h}$-modules. (There is no need to "start the induction", since we are assuming nothing in the case where there is no μ with precisely h non-zero parts such that $\mu \rhd \lambda$.)

Now, $M_\lambda E_h \cong M_{\hat{\lambda}}$ (as a $\bar{K}G_{n-h}$-module), by Corollary 10.15. Act on the series in (17.4) by E_h:

$$M_{\hat{\lambda}} \cong M_\lambda E_h = M_0 E_h \supseteq M_1 E_h \supseteq \ldots \supseteq M_t E_h \supseteq M_{t+1} E_h = 0.$$

This is a series of $\bar{K}G_{n-h}$-submodules of $M_{\hat{\lambda}}$, and for $0 \le s \le t-1$,

$M_sE_h/M_{s+1}E_h \cong S_{\hat{\mu}}$, if $M_s/M_{s+1} \cong S_\mu$ and μ has h non-zero parts,

$M_sE_h = M_{s+1}E_h$, if $M_s/M_{s+1} \cong S_\mu$ and μ has fewer than h non-zero parts.

In particular,

(17.7) No composition factor of $M_\lambda E_h/S_\lambda E_h = M_0E_h/M_tE_h$ is isomorphic to $D_{\hat{\lambda}}$.

Also, by (17.6), $\dim(M_\lambda E_h/S_\lambda E_h) = \dim(M_{\hat{\lambda}}/\hat{M}_u)$, so

(17.8) $\dim(S_\lambda E_h) = \dim S_{\hat{\lambda}}$.

Let θ denote our $\bar{K}G_{n-h}$-isomorphism from $M_{\hat{\lambda}}$ to $M_\lambda E_h$. By the Submodule Theorem, either $S_\lambda E_h \supseteq S_{\hat{\lambda}}\theta$ or $S_\lambda E_h \subseteq S_{\hat{\lambda}}^\perp\theta$. But a composition factor of $(M_{\hat{\lambda}}\theta)/(S_{\hat{\lambda}}^\perp\theta)$ is isomorphic to $D_{\hat{\lambda}}$, so by (17.7) $S_\lambda E_h \nsubseteq S_{\hat{\lambda}}^\perp\theta$. Hence, using (17.8) we have

$$S_\lambda E_h = S_{\hat{\lambda}}\theta,$$

and this proves the first part of the theorem.

We next wish to show that $S_\lambda^\perp E_h = S_{\hat{\lambda}}^\perp\theta$. The proof of this is made awkward by the fact that the bilinear form on M_λ (see Definition 11.1) is usually singular when restricted to $M_\lambda E_h$.

Let us recall the definition of the $\bar{K}G_{n-h}$-isomorphism θ from $M_{\hat{\lambda}}$ to $M_\lambda E_h$. First of all, in Theorem 10.14, we constructed an injection ϕ from $M_{\hat{\lambda}}$ into M_λ. This injection clearly has the property that

$$\langle \xi_1, \xi_2 \rangle_{\hat{\lambda}} = \langle \xi_1\phi, \xi_2\phi \rangle_\lambda \text{ for all } \xi_1, \xi_2 \in M_{\hat{\lambda}}.$$

Then θ is defined by

$$\xi\theta = \xi\phi E_h \qquad (\xi \in M_{\hat{\lambda}}).$$

Now,

$$E_h = \frac{1}{q^{|\Gamma(h)|}} \sum_{g \in G(\Gamma(h))} \chi_{c_h}(g)g.$$

Let

$$\bar{E}_h = \frac{1}{q^{|\Gamma(h)|}} \sum_{g \in G(\Gamma(h))} \chi_{c_h}^{-1}(g)g,$$

and define the map $\bar{\theta}$ from $M_{\hat{\lambda}}$ into $M_\lambda \bar{E}_h$ by

$$\xi\bar{\theta} = \xi\phi\bar{E}_h \qquad (\xi \in M_{\hat{\lambda}}).$$

Using (10.10), we find that

$$q^{|\Gamma_2|} \langle \xi_1\theta, \xi_2\bar{\theta}\rangle_\lambda = q^{|\Gamma_2|} \langle \xi_1\phi E_h, \xi_2\phi\bar{E}_h\rangle_\lambda = \langle \xi_1\phi, \xi_2\phi\rangle_\lambda$$

$$= \langle \xi_1, \xi_2\rangle_{\hat{\lambda}}, \text{ for all } \xi_1, \xi_2 \in M_{\hat{\lambda}}.$$

where $\Gamma_2 = \{(i, j) \mid i > j \leq h \text{ and } \mathrm{row}_{T_\lambda\pi_R}(i) > \mathrm{row}_{T_\lambda\pi_R}(j)\}$,

and R is the unique element of $\mathcal{R}_h$.

Now, for all $\zeta \in S_\lambda^\perp$, there exists $\xi \in M_{\hat{\lambda}}$ with $\zeta E_h = \xi\theta$ (because $M_\lambda E_h$ $M_{\hat{\lambda}}\theta$), and for all $\eta \in S_{\hat{\lambda}}$ there exists $\eta' \in S_\lambda$ with $\eta'\bar{E}_h = \eta\bar{\theta}$ (because $S_\lambda\bar{E}_h$ $S_{\hat{\lambda}}\bar{\theta}$, by analogy with the result we have already proved). Therefore,

$$q^{-|\Gamma_2|} \langle \xi, \eta\rangle_{\hat{\lambda}} = \langle \zeta E_h, \eta'\bar{E}_h\rangle_\lambda = \langle \zeta, \eta'\bar{E}_h{}^2\rangle_\lambda,$$

which equals zero, since $\zeta \in S_\lambda^\perp$ and $\eta'\bar{E}_h{}^2 \in S_\lambda$. Hence $\xi \in S_{\hat{\lambda}}^\perp$, and we have proved that

$$S_\lambda^\perp E_h \subseteq S_{\hat{\lambda}}^\perp\theta.$$

On the other hand, $S_\lambda^\perp$ has the same $\bar{K}G_n$ composition factors as M_λ/S_λ, so $S_\lambda^\perp E_h$ has the same $\bar{K}G_{n-h}$ composition factors as $M_\lambda E_h/S_\lambda E_h$. Thus,

$$\dim(S_\lambda^\perp E_h) = \dim M_{\hat{\lambda}} - \dim S_{\hat{\lambda}} = \dim(S_{\hat{\lambda}}^\perp\theta).$$

Therefore,

$$S_\lambda^\perp E_h = S_{\hat{\lambda}}^\perp\theta.$$

Since E_h is an idempotent element,

$$(S_\lambda \cap S_\lambda^\perp)E_h = (S_\lambda E_h) \cap (S_\lambda^\perp E_h) = (S_{\hat{\lambda}}\theta) \cap (S_{\hat{\lambda}}^\perp\theta)$$

$$= (S_{\hat{\lambda}} \cap S_{\hat{\lambda}}^\perp)\theta.$$

herefore,

$$D_\lambda E_h = S_\lambda E_h/(S_\lambda \cap S_\lambda^\perp)E_h = S_{\hat\lambda}\theta/(S_{\hat\lambda} \cap S_{\hat\lambda}^\perp)\theta$$

$$\cong S_{\hat\lambda}/(S_{\hat\lambda} \cap S_{\hat\lambda}^\perp), \text{ a } \bar{K}G_{n-h}\text{-isomorphism,}$$

$$= D_{\hat\lambda}.$$

his is what we wished to prove. ■

Now, assume that M is a $\bar{K}G_n$-module with a composition series

$$M = M_0 \supset M_1 \supset \ldots \supset M_k \supset M_{k+1} = 0,$$

hich has the property that, for each s, M_s/M_{s+1} is isomorphic to $D_{\mu^{(s)}}$ here $\mu^{(s)}$ is a partition of n into at most h non-zero parts. Then

$$ME_h = M_0E_h \supseteq M_1E_h \supseteq \ldots \supseteq M_kE_h \supseteq M_{k+1}E_h = 0$$

s a series of $\bar{K}G_{n-h}$-submodules of ME_h, and by Theorem 17.3 and (17.2),

$$M_sE_h/M_{s+1}E_h \cong D_{\hat\mu^{(s)}} \text{ if } \mu^{(s)} \text{ has } h \text{ non-zero parts,}$$

$$M_sE_h = M_{s+1}E_h \text{ if } \mu^{(s)} \text{ has fewer than } h \text{ non-zero parts.}$$

f we take $M = S_\lambda$, then Theorem 17.1 follows at once, since $S_\lambda E_h \cong S_{\hat\lambda}$.

There is a highly "symmetric" generator for S_λ when λ has precisely two non-zero parts. We could avoid using this generator later on, but the construction of it is particularly elegant, and the power of the Submodule Theorem and the Kernel Intersection Theorem is well illustrated at a critic point.

We shall assume that $\lambda = (n - m, m)$ with $n - m \geq m > 0$, and regard M as the vector space over $\bar{K}$ with basis elements indexed by the m-dimensional subspaces of V. The generator of S_λ described in the second equality of 11.12(i) involves

$$q^{|\Gamma|} = q^{(m+1)m/2}$$

subspaces of V. In Example 11.17, we gave a generator using only

$$2^m q^{m(m-1)/2} = q^{(m+1)m/2} (2/q)^m$$

subspaces of V. The generator we shall construct in this section involves

$$\prod_{i=0}^{m-1} (q^i + 1) = 2(q + 1)(q^2 + 1) \ldots (q^{m-1} + 1)$$

subspaces of V, which is a smaller number still, and we feel confident that no non-zero element of S_λ involves fewer subspaces. The standard generator for the corresponding $\mathfrak{S}_n$-module involves 2^m terms.

18.1 DEFINITIONS

(i) Let V* be an arbitrary subspace of V of dimension 2m, and let $f_1, f_2, \ldots, f_m, g_1, g_2, \ldots, g_m$ be a basis for V*.

(ii) Define a bilinear form (,) on V* by

$$(\alpha_1 f_1 + \ldots + \alpha_m f_m + \beta_1 g_1 + \ldots + \beta_m g_m, \alpha_1' f_1 + \ldots + \alpha_m' f_m + \beta_1' g_1 + \ldots + \beta_m' g_m)$$

$$= \sum_{i=1}^{m} (\alpha_i \beta_i' + \alpha_i' \beta_i) \qquad (\alpha_i, \beta_i, \alpha_i', \beta_i' \in \mathbb{F}_q).$$

(iii) Define the function ϕ from V^* to $\mathbb{F}_q$ by

$$(\alpha_1 f_1 + \dots + \alpha_m f_m + \beta_1 g_1 + \dots + \beta_m g_m)\phi = \alpha_1\beta_1 + \dots + \alpha_m\beta_m.$$

(iv) If W is a subspace of V^*, let

$$W^{\perp} = \{v \in V^* \mid (v, w) = 0 \text{ for all } w \in w\},$$

$$\text{rad}\, W = \{w \in W \mid w\phi = 0 \text{ and } (w, w') = 0 \text{ for all } w' \in W\}.$$

(v) A subspace W of V^* is said to be <u>isotropic</u> if $W = \text{rad}\, W$.

The Gram matrix for the bilinear form (,) with respect to the chosen basis of V^* is

$$\left(\begin{array}{c|c} 0 & I_m \\ \hline I_m & 0 \end{array}\right).$$

One readily verifies that for $v_1, v_2 \in V$, $\alpha, \beta \in \mathbb{F}_q$, we have

(18.2) $\quad (\alpha v_1 + \beta v_2)\phi = \alpha^2(v_1\phi) + \beta^2(v_2\phi) + \alpha\beta(v_1, v_2)$, and

$$(v_1, v_1) = 2(v_1\phi).$$

The function ϕ has been introduced only to cope with the case where q is a power of 2.

If W is a subspace of V^*, then (18.2) shows that $W^{\perp}$ and rad W are subspaces of V^*, and rad $W \subseteq W \cap W^{\perp}$ with equality if q is odd.

Although the next four results are known, we supply the proofs, since they are short and because they are normally given only for the case where q is odd. (Our source is Lang [L].)

18.3 LEMMA. <u>If v and w are linearly independent elements of V^* with $(v, w) \neq 0$ and $w\phi = 0$, then there exists $v_1 \in V^*$ such that</u>

(i) *v_1 is a linear combination of v and w, but v_1 is independent of w.*

(ii) *$(v_1, w) = 1$*

(iii) *$v_1\phi = 0$*

Proof: Replacing v be a scalar multiple of itself, we may assume that (v, w) = 1. Since $w\phi = 0$, we have $(w, w) = 0$. Let $\alpha = -v\phi \in \mathbb{F}_q$ and $v_1 = \alpha w + v$. Then our results follow, because

$$(\alpha w + v, w) = 1 \quad \text{and} \quad (\alpha w + v)\phi = v\phi + \alpha = 0.$$

■

18.4 LEMMA. *Suppose that W is a subspace of V* and let U be a complementary subspace to rad W in W. Let $w_1, \ldots, w_s$ be a basis for rad W. Then there exist $v_1, \ldots, v_s$ such that*

(i) *For every non-zero element u of U, $u, w_1, \ldots, w_s, v_1, \ldots, v_s$ are linearly independent.*

(ii) $(u, v_i) = 0$ *for all* $u \in U$, $1 \leq i \leq s$.

(iii) $(v_i, v_j) = v_i\phi = 0$ *for* $1 \leq i, j \leq s$.

(iv) $(w_i, v_j) = 1$ *if* $i = j$

$= 0$ *if* $i \neq j$ *for* $1 \leq i, j \leq s$.

Proof: Let $U_1 = \langle w_2, \ldots, w_s\rangle \oplus U$. Then $U_1 \subset W$, so $W^\perp \subset U_1^\perp$, and there exists $v \in U_1^\perp \setminus W^\perp$. Thus, $(v, w_1) \neq 0$, and Lemma 18.3 gives an element v_1 such that

$$(v_1, w_j) = 0 \text{ for } j > 1, \quad (v_1, w_1) = 1 \text{ (so } v_1 \notin W), \quad \text{and } v_1\phi = 0.$$

Now let $W_1 = \langle w_1, \ldots, w_s\rangle \oplus \langle v_1\rangle \oplus U$. Then rad $W_1 = \langle w_2, \ldots, w_s\rangle$, and we may continue inductively to construct $v_2, \ldots, v_s$.

■

18.5 COROLLARY. *If W is an isotropic subspace of V*, then dim W ≦ m.*

Proof: This is an immediate consequence of conclusion (i) of the Lemma.

■

8.6 COROLLARY. *Suppose that W is an m-dimensional isotropic subspace of* **, and let $w_1, \ldots, w_m$ be a basis of W. Then there exists a basis $w_1, \ldots,$* *m, $v_1, \ldots, v_m$ of V* such that*

$$(w_i, w_j) = (v_i, v_j) = w_i\phi = v_i\phi = 0$$

$$(w_i, v_j) = 1 \quad \text{if } i = j$$

$$= 0 \quad \text{if } i \neq j \quad \text{for } 1 \leq i, j \leq m.$$

■

his says that we may extend any basis of an m-dimensional isotropic space o a basis of V* in such a way that the Gram matrix takes the form

$$\left(\begin{array}{c|c} 0 & I_m \\ \hline I_m & 0 \end{array}\right).$$

Our next main objective is to prove:

8.7 THEOREM. *Suppose that W_1, W_2, W_3 are isotropic subspaces of V*, each* *f dimension m. Then*

$$m - \dim (W_1 \cap W_2) - \dim (W_2 \cap W_3) - \dim (W_3 \cap W_1)$$

s even.

There is a neat proof of this theorem when q is odd. For it can be hown that if W_2 and W_3 are m-dimensional isotropic spaces, with dim $(W_2 \cap W_3$ r_1, then every isometry θ_1 sending W_2 onto W_3 has determinant $(-1)^{m-r_1}$. hus, if dim $(W_1 \cap W_3) = r_2$, dim $(W_1 \cap W_2) = r_3$, and θ_2 is an isometry from W_3 to W_1, then $\theta_1\theta_2$ is an isometry from W_2 to W_1, and

$$(-1)^{m-r_1}(-1)^{m-r_2} = \det \theta_1 \det \theta_2 = \det \theta_1\theta_2 = (-1)^{m-r_3}.$$

his proves Theorem 18.7 in the case where q is odd. Apparently though, more ork must be done to prove the theorem in the general case; we will be ewarded, however, by other useful results on the way.

18.8 LEMMA. *Suppose that W_1, W_2 are m-dimensional subspaces of V^*, with W_1 isotropic and $\dim(W_1 \cap W_2) = r$. Let $w_1, \ldots, w_m$ be a basis of W_1 containing a basis $w_1, \ldots, w_r$ of $W_1 \cap W_2$, and let $w_1, \ldots, w_m, v_1, \ldots, v_m$ be a basis of V^* having the properties described in Corollary 18.6. Then W_2 is an isotropic subspace if and only if W_2 has a basis of the form*

$$w_1, \ldots, w_r, x_{r+1}, \ldots, x_m$$

where

$$x_i = \sum_{j=r+1}^{m} \alpha_{ij} w_j + v_i \qquad (\alpha_{ij} \in \mathbb{F}_q, \ r+1 \le i \le m),$$

with

$$\alpha_{ii} = 0, \quad \alpha_{ij} = -\alpha_{ji} \qquad (r+1 \le i, j \le m).$$

Proof: There is a basis for W_2 of the form

$$w_1, \ldots, w_r, y_{r+1}, \ldots, y_m$$

where

$$y_i = \sum_{j=r+1}^{m} \alpha_{ij} w_j + \sum_{j=1}^{m} \beta_{ij} v_j \qquad (\alpha_{ij}, \beta_{ij} \in \mathbb{F}_q, \ r+1 \le i \le m).$$

Suppose that W_2 is isotropic. Then

$$y_i = \sum_{j=r+1}^{m} (\alpha_{ij} w_j + \beta_{ij} v_j) \qquad \text{for } r+1 \le i \le m,$$

since $(w_i, y_k) = 0$ for $1 \le i \le r < k \le m$. But $w_1, \ldots, w_m, y_{r+1}, \ldots, y_m$ are linearly independent, so the matrix

$$(\beta_{ij})_{r+1 \le i, j \le m}$$

is non-singular. Therefore, we may assume that

$$y_i = \sum_{j=r+1}^{m} \alpha_{ij} w_j + v_i \qquad \text{for } r+1 \le i \le m.$$

Since $y_i\phi = (y_i, y_i) = 0$, we obtain

$$\alpha_{ii} = 0, \qquad \alpha_{ij} = -\alpha_{ji} \qquad (r + 1 \leq i, j \leq m).$$

Conversely, if these conditions hold, then W_2 is isotropic.

■

18.9 COROLLARY. _Suppose that W_1 is an m-dimensional isotropic subspace of V^* and U is an r-dimensional subspace of W_1. Then the number of m-dimensional isotropic subspaces W_2 such that $W_1 \cap W_2 = U$ is_

$$q^{(m-r)(m-r-1)/2}.$$

Proof: $q^{(m-r)(m-r-1)/2}$ is the number of choices for the matrix

$$(\alpha_{ij})_{r+1 \leq i,j \leq m}$$

in the statement of the lemma.

■

18.10 COROLLARY. _The number of m-dimensional isotropic subspaces of V^* is_

$$\prod_{i=0}^{m-1} (q^i + 1).$$

Proof: $W_1 = \langle f_1, f_2, \ldots, f_m \rangle$ is an m-dimensional isotropic subspace of V^*. For each r with $0 \leq r \leq m$, the number of r-dimensional subspaces of W_1 is

$$\begin{bmatrix} m \\ r \end{bmatrix}.$$

Therefore, using the last corollary, the number of isotropic subspaces of V^* is

$$\sum_{r=0}^{m} \begin{bmatrix} m \\ r \end{bmatrix} q^{(m-r)(m-r-1)/2} = \prod_{i=0}^{m-1} (q^i + 1),$$

the equality being obtained by putting x = 1 in Theorem 3.5.

■

Proof of Theorem 18.7: Suppose we have three isotropic subspaces W_1, W_2, W_3 of V^*, each of dimension m. Let

$$d = \dim (W_1 \cap W_2 \cap W_3)$$

$d + c = \dim (W_1 \cap W_2)$

$d + b = \dim (W_1 \cap W_3)$.

Choose a basis $w_1, w_2, \ldots, w_m$ for W_1 such that

$w_1, \ldots, w_d$ is a basis for $W_1 \cap W_2 \cap W_3$

$w_1, \ldots, w_{d+c}$ is a basis for $W_1 \cap W_2$

$w_1, \ldots, w_d, w_{d+c+1}, \ldots, w_{d+c+b}$ is a basis for $W_1 \cap W_3$.

Construct a basis $w_1, \ldots, w_m, v_1, \ldots, v_m$ of V^* satisfying the conclusion of Corollary 18.6. Take a basis for W_2, as in Lemma 18.8, of the form

$w_1, \ldots, w_{d+c}, y_{d+c+1}, \ldots, y_m$

where

$$y_i = \sum_{j=d+c+1}^{m} \beta_{ij} w_j + v_i \qquad (\beta_{ij} \in \mathbb{F}_q,\ d + c + 1 \leq i \leq m).$$

Apply Lemma 18.8 again, to choose a basis of W_3:

$w_1, \ldots, w_d, z_{d+1}, \ldots, z_{d+c}, w_{d+c+1}, \ldots, w_{d+c+b}, z_{d+c+b+1}, \ldots, z_m,$

where

$$z_i = \sum_{j=d+1}^{d+c} \gamma_{ij} w_j + \sum_{j=d+c+b+1}^{m} \gamma_{ij} w_j + v_i$$

$(\gamma_{ij} \in \mathbb{F}_q,\ i \in \{d + 1, \ldots, d + c\} \cup \{d + c + b + 1, \ldots, m\})$.

Then $W_2 + W_3$ is spanned by

$$\{w_1, \ldots, w_{d+c+b}\}$$

$$\cup \{ \sum_{j=d+c+b+1}^{m} \beta_{ij} w_j + v_i \mid d + c + 1 \leq i \leq m\}$$

$$\cup \{ \sum_{j=d+c+b+1}^{m} \gamma_{ij} w_j + v_i \mid d + 1 \leq i \leq d + c\}$$

$$\cup \{ \sum_{j=d+c+b+1}^{m} \gamma_{ij} w_j + v_i \mid d + c + b + 1 \le i \le m \}.$$

Hence $W_2 + W_3$ is spanned by the first three of these sets together with

$$\{ \sum_{j=d+c+b+1}^{m} (\beta_{ij} - \gamma_{ij}) w_j \mid d + c + b + 1 \le i \le m \}.$$

The number of linearly independent elements in this last set is even, since the matrix

$$(\beta_{ij} - \gamma_{ij})_{d+c+b+1 \le i,j \le m}$$

is alternating. Therefore,

$$\begin{aligned} \dim (W_2 + W_3) &\equiv (d + c + b) + (m - d - c) + c \\ &\equiv b + c + m \qquad \text{(modulo 2)}. \end{aligned}$$

Thus,

$$\begin{aligned} \dim (W_1 \cap W_2) + \dim (W_2 \cap W_3) &+ \dim (W_3 \cap W_1) \\ &\equiv (d + c) + (b + c + m) + (d + b) \\ &\equiv m \qquad \text{(modulo 2)}, \end{aligned}$$

as stated in Theorem 18.7.

■

18.11 DEFINITION. Choose any m-dimensional isotropic subspace J of V*. For each m-dimensional isotropic subspace W of V*, define sign W ∈ K by

$$\text{sign } W = (-1_K)^{\dim(J \cap W)}.$$

Theorem 18.7 shows that for any two m-dimensional isotropic subspaces W_1, W_2 of V*, we have

$$\text{(18.12)} \quad \text{sign } W_1 \cdot \text{sign } W_2 = (-1)^{m - \dim(W_1 \cap W_2)}.$$

In particular, the ratio of sign W_1 to sign W_2 is independent of the isotropic space J chosen in the definition: that is, if we alter J, then the signs of the isotropic spaces either all remain the same, or they all change.

Consider the element

$$\sum_{W} (\text{sign } W)W$$

of $M_{(n-m,m)}$, the sum being over all m-dimensional isotropic subspaces W of V*. We shall employ the Kernel Intersection Theorem and the Submodule Theorem, respectively, to prove that this element belongs to, and indeed generates, $S_{(n-m,m)}$. The pictures (2.7) and (2.8) illustrate this generator in the case where m = 2.

18.13 THEOREM. $\underline{\sum_{W} (\text{sign } W)W \text{ belongs to } S_{(n-m,m)}}$.

<u>Proof</u>: Suppose that $0 \le i \le m - 1$, and let U be an i-dimensional subspace of V* contained in some m-dimensional isotropic subspace W_1 of V*. The number of r-dimensional subspaces of W_1 which contain U is

$$\begin{bmatrix} m - i \\ m - r \end{bmatrix},$$

by Corollary 3.3(i). Each of these r-dimensional spaces is automatically isotropic. Using (18.12) and Corollary 18.9, we have

$$\Sigma\{\text{sign } W_2 \mid W_2 \text{ is an m-dimensional isotropic subspace of V* and } W_2 \supseteq U$$

$$= \text{sign } W_1 \sum_{r=i}^{m} (-1)^{m-r} \begin{bmatrix} m - i \\ m - r \end{bmatrix} q^{(m-r)(m-r-1)/2}$$

$$= \text{sign } W_1 \sum_{j=0}^{m-i} (-1)^{j} \begin{bmatrix} m - i \\ j \end{bmatrix} q^{j(j-1)/2} .$$

Since $m - i > 0$, we find that the last sum is zero, by putting $x = -1$ in Theorem 3.5.

We have proved that

$$\sum_{W} (\text{sign } W)W \in \text{Ker } \psi_{1,i} \quad \text{when } 0 \le i \le m - 1,$$

and the result now follows from the Kernel Intersection Theorem 13.3.

■

18.14 THEOREM. $\sum_W$ (sign W)W generates $S_{(n-m,m)}$ as a KG_n-module.

Proof: The generator ξ of $S_{(n-m,m)}$ described in Example 11.17(v) involves only subspaces of $\langle e_1, e_2, \ldots, e_{2m}\rangle$, so we take this 2m-dimensional space to be V^*, and let

$$f_i = e_{2i-1}$$

$$g_i = e_{2i} \qquad (1 \le i \le m)$$

in Definition 18.1. Every space which occurs in ξ has the form

$$\begin{array}{llll}\langle\alpha_{11}f_1 + g_1, & & & \\ \alpha_{21}f_1 & + \alpha_{23}f_2 + g_2, & & \\ \alpha_{31}f_1 & + \alpha_{33}f_2 & + \alpha_{35}f_3 + g_3, & \\ \vdots & \vdots & \vdots & \\ \alpha_{m1}f_1 & + \alpha_{m3}f_2 & + \quad \cdots & + \alpha_{m,2m-1}f_m + g_m\rangle,\end{array}$$

with $\alpha_{ij} \in \mathbb{F}_q$. The definition of the bilinear form on V^* ensures that the space we have described is isotropic if and only if $\alpha_{ij} = 0$ for all i, j. Therefore,

$$\langle \xi, \sum_W (\text{sign } W)W\rangle_\lambda = \text{sign } (\langle g_1, g_2, \ldots, g_m\rangle) \neq 0,$$

which proves that $\sum_W$ (sign W)W $\in S_{(n-m,m)} \setminus S^{\perp}_{(n-m,m)}$. The KG_n-module generated by our element must then equal $S_{(n-m,m)}$, by the Submodule Theorem.

■

The form of the Kernel Intersection Theorem means that the problem of deciding whether or not a particular element of M_λ belongs to S_λ often reduces to one of determining which Gaussian polynomials are divisible by the characteristic of K - Theorem 16.12 illustrates this point. Indeed, in the next section we shall determine all the composition factors of S_λ, when λ has just two parts, using techniques which rely heavily on the information we glean about Gaussian polynomials in the present chapter.

19.1 DEFINITIONS. Suppose that p is a prime number.

(i) If b is a positive integer, let $\nu_p(b)$ be the largest integer i such that p^i divides b.

(ii) If b is a non-negative integer, let $\ell_p(b)$ be the least non-negative integer i such that $b < p^i$.

For the results we prove here, it is sufficient to assume that q is a positive integer coprime to p, so let this be the case throughout the chapter.

19.2 LEMMA. <u>If p is an odd prime, q > 1, and s, t are positive integers such that p divides $(q^s - 1)$, then</u>

$$\nu_p\left(\frac{q^{st} - 1}{q^s - 1}\right) = \nu_p(t).$$

<u>Proof</u>: If $q^s - 1 = ap^b$, with a coprime to p, then

$$\frac{q^{sp} - 1}{q^s - 1} = \frac{(ap^b + 1)^p - 1}{ap^b} = \sum_{i=1}^{p} \binom{p}{i} (ap^b)^{i-1}$$

$$\equiv p \qquad (\text{modulo } p^2), \text{ since } p > 2.$$

Suppose that $t = cp^d$, with c coprime to p. Then

$$\frac{q^{st} - 1}{q^s - 1} = \frac{q^{sc} - 1}{q^s - 1} \prod_{i=1}^{d} \left(\frac{q^{scp^i} - 1}{q^{scp^{i-1}} - 1} \right)$$

But

$$\frac{q^{sc} - 1}{q^s - 1} = 1 + q^s + \ldots + q^{s(c-1)} \equiv c \qquad (\text{modulo } p),$$

so

$$\nu_p \left(\frac{q^{st} - 1}{q^s - 1} \right) = \sum_{i=1}^{d} \nu_p \left(\frac{q^{scp^i} - 1}{q^{scp^{i-1}} - 1} \right) = \sum_{i=1}^{d} 1$$

$$= d = \nu_p(t).$$

■

19.3 LEMMA. <u>Let p be a prime number.</u>

(i) <u>If p divides $(q - 1)$, then</u> $\begin{bmatrix} n \\ m \end{bmatrix} \equiv \binom{n}{m}$ <u>modulo p.</u>

(ii) <u>If</u> $n = n_0 + n_1 p + \ldots + n_k p^k$, <u>and</u>

$$m = m_0 + m_1 p + \ldots + m_k p^k \qquad (0 \le n_i, m_i < p),$$

<u>then</u>

$$\binom{n}{m} \equiv \binom{n_0}{m_0} \binom{n_1}{m_1} \ldots \binom{n_k}{m_k} \quad \text{modulo } p.$$

<u>In particular, p divides</u> $\binom{n}{m}$ <u>if and only if $n_i < m_i$ for some i.</u>

<u>Proof</u>: Since

$$(1 + x)(1 + xq) \ldots (1 + xq^{n-1}) = \sum_{i=0}^{m} \begin{bmatrix} n \\ i \end{bmatrix} q^{i(i-1)/2} x^i,$$

we get

$$\sum_{i=0}^{n} \begin{bmatrix} n \\ i \end{bmatrix} x^i \equiv (1 + x)^n \quad \text{if } p \text{ divides } (q - 1)$$

$$= \sum_{i=0}^{n} \binom{n}{i} x^i$$

$$\equiv (1 + x)^{n_0}(1 + x^p)^{n_1} \ldots (1 + x^{p^k})^{n_k} \quad \text{modulo } p.$$

All the results follow by considering the coefficient of x^m.

■

19.4 DEFINITIONS

(i) Let e be the least positive integer such that p divides $(q^e - 1)$

(ii) If b is a non-negative integer, define b* by

$b = b^*e + b' \quad 0 \le b' < e.$

Thus, e is the exponent of q modulo p, and b* is the integral part of b/e.

19.5 THEOREM.

<u>Let a and b be positive integers. Then the prime p divides each of</u>

$$\begin{bmatrix} a \\ 1 \end{bmatrix}, \begin{bmatrix} a+1 \\ 2 \end{bmatrix}, \begin{bmatrix} a+2 \\ 3 \end{bmatrix}, \ldots, \begin{bmatrix} a+b-1 \\ b \end{bmatrix}$$

<u>if and only if both the following conditions hold</u>:

(i) <u>e divides a, so a = a*e</u>.

(ii) <u>$a^* \equiv 0 \quad \text{modulo } p^{\ell_p(b^*)}$</u>.

We remark that condition (ii) holds, trivially, if b* = 0.

<u>Proof</u>: First note that p divides each of

$$\binom{a^*}{1}, \binom{a^*+1}{2}, \ldots, \binom{a^*+b^*-1}{b^*}$$

if and only if p divides each of

$$\binom{a^*}{1}, \binom{a^*}{2}, \ldots, \binom{a^*}{b^*},$$

(consider Pascal's Triangle), and this happens if and only if

$$a^* \equiv 0 \mod p^{\ell_p(b^*)},$$

by applying the last part of Lemma 19.3.

If $e = 1$, then p divides $(q - 1)$, and the theorem follows immediately from Lemma 19.3(i). Assume, therefore, that $e > 1$. Then p is odd, $q > 1$, and

$$p \text{ divides } \begin{bmatrix} a \\ 1 \end{bmatrix} = \frac{q^a - 1}{q - 1} \Leftrightarrow p \text{ divides } (q^a - 1)$$

$$\Leftrightarrow e \text{ divides } a.$$

We may thus assume that $a = a^*e$. Consider

$$\begin{bmatrix} a + i - 1 \\ i \end{bmatrix} = \frac{(q^{a+i-1} - 1)(q^{a+i-2} - 1) \ldots (q^a - 1)}{(q^i - 1) \,.\, (q^{i-1} - 1) \ldots (q - 1)} .$$

Let $i = i^*e + i'$. Then $\{a + i - 1, a + i - 2, \ldots, a\}$ contains i^* integers divisible by e if $i' = 0$, and the set contains $(i^* + 1)$ integers divisible by e if $i' \geq 1$. Therefore,

$$\nu_p \begin{bmatrix} a + i - 1 \\ i \end{bmatrix} = \nu_p \begin{pmatrix} a^* + i^* - 1 \\ i^* \end{pmatrix}, \text{ if } i' = 0$$

$$\nu_p \begin{bmatrix} a + i - 1 \\ i \end{bmatrix} > 0, \text{ if } i' \geq 1,$$

by Lemma 19.2. Thus, p divides each of

$$\begin{bmatrix} a \\ 1 \end{bmatrix}, \begin{bmatrix} a + 1 \\ 2 \end{bmatrix}, \ldots, \begin{bmatrix} a + b - 1 \\ b \end{bmatrix}$$

if and only if e divides a, and p divides each of

$$\begin{pmatrix} a^* \\ 1 \end{pmatrix}, \begin{pmatrix} a^* + 1 \\ 2 \end{pmatrix}, \ldots, \begin{pmatrix} a^* + b^* - 1 \\ b^* \end{pmatrix}.$$

Again, the theorem follows from the first paragraph of the proof. ∎

19.6 THEOREM. _Let $\mu = (\mu_1, \mu_2, \ldots, \mu_h)$ be a partition of n, with $\mu_h > 0$. Then the trivial $\bar{K}G_n$-module is a submodule of S_μ if and only if_

(i) e divides $(\mu_d + 1)$, and

(ii) $(\mu_d + 1)^* \equiv 0$ modulo $p^{\ell_p(\mu^*_{d+1})}$

<u>for each d with $1 \leq d \leq h - 1$.</u>

<u>Proof</u>: This follows immediately from Theorems 16.12 and 19.5. ■

Condition (i) of the above theorem guarantees that the diagrams $[\mu]$ and $[n]$ have the same e-core, in agreement with the Theorem 1.2 of Fong and Srinivasan, although this is not obvious. It is interesting to compare the theorem with the corresponding situation for $\mathfrak{S}_n$, where the relevant condition is

$$\mu_d + 1 \equiv 0 \quad \text{modulo } p^{\ell_p(\mu_{d+1})} \qquad (1 \leq d \leq h - 1).$$

This is equivalent to conditions (i) and (ii), when $q = 1$, for then p divides $q - 1$, so $e = 1$.

The remaining results in this section are purely numerical. The integer q is not mentioned again, and we may take p to be any integer greater than 1, and e to be an arbitrary positive integer, although when we apply the results, q will be a prime power, p a prime number not dividing q, and e will be the exponent of q modulo p. We continue to denote the integral part of (b/e) by b*.

19.7 DEFINITIONS. Let $p \geq 2$, $e \geq 1$ be integers.

(i) Given two non-negative integers, a and b, we say that <u>a contains b to base p</u> if and only if $a > 0$,

$$a = a_0 + a_1p + \dots + a_kp^k \qquad (0 \leq a_i < p,\ a_k \neq 0),$$

$$b = b_0 + b_1p + \dots \qquad (0 \leq b_i < p),$$

$b_k = b_{k+1} = \dots = 0$, and for each i, either $b_i = 0$ or $b_i = a_i$.

(ii) Given two non-negative integers, n and m, let

$f_{p,e}(n, m) = 1$, if $(n + 1)^*$ contains m^* to base p, and either e divides m or e divides $(n + 1 - m)$,

$f_{p,e}(n, m) = 1$, if $m = 0$,

$f_{p,e}(n, m) = 0$, otherwise.

A few remarks about these definitions are in order. Note that in Definition 19.7(i) we require that $b_k = 0$; hence if a contains b to base p, then $a \geq 2b + 1$. It is easy to verify from this that $f_{p,e}(n, m) = 1$ only if $n \geq 2m$.

The statement that $f_{p,e}(n, m) = 1$, if $m = 0$, in Definition 19.7(ii) is not redundant; consider the case $n + 1 < e$, $m = 0$. The numbers m, $(n + 1 - m)$ appearing in the definition of $f_{p,e}(n, m)$ are the "first column hook lengths" (see Definition 20.1) for the diagram $[n - m, m]$, provided that $n - m \geq m > 0$. We introduce the function $f_{p,e}$ since we shall prove in the next chapter that $f_{p,e}(n, m)$ is the composition multiplicity of the trivial $\bar{K}G_n$-module in $S_{(n-m,m)}$, when $n - m \geq m$, K has the characteristic p, and e is the exponent of q modulo p.

We consider $f_{p,e}(n - 1, m) + f_{p,e}(n - 1, m - 1)$ in the next two Lemmas, as this number enables us to apply, inductively, results for G_{n-1}.

19.8 LEMMA. Assume that $m \geq 1$ and $f_{p,e}(n, m) = 1$. Then $f_{p,e}(n - 1, m) + f_{p,e}(n - 1, m - 1) = 1$.

Proof: Note first that $n - 1$ is positive since $f_{p,e}(n, m) = 1$ implies that $n \geq 2m$.

Hereafter, we adopt the notation $a \mid b$ for "a divides b" and $a \supset b$ for "a contains b to base p".

Since $m \geq 1$, we have $(n + 1)^* \supset m^*$, and $e \mid m$ or $e \mid (n + 1 - m)$.

If $e \mid m$ and $e \nmid (n + 1 - m)$, then $e \neq 1$, so $e \nmid (m - 1) \geq 1$. Therefore, $f_{p,e}(n - 1, m - 1) = 0$. Also, $(n + 1)^* = n^*$, so $n^* \supset m^*$, and $f_{p,e}(n - 1, m) = 1$.

If $e \nmid m$ and $e \mid (n + 1 - m)$, then $e \neq 1$, so $e \mid (n - m)$. Therefore, $f_{p,e}(n - 1, m) = 0$. Also, $(n + 1)^* = n^*$ and $m^* = (m - 1)^*$, so $n^* \supset (m - 1)^*$ and $f_{p,e}(n - 1, m - 1) = 1$.

If $e \mid m$ and $e \mid (n + 1 - m)$, then $(m - 1)^* = m^* - 1$ and $n^* + 1 = (n + 1)^*$. Since $(n + 1)^* \supset m^*$,

$$n^* + 1 = a_j p^j + a_{j+1} p^{j+1} + \dots$$

$$m^* = b_j p^j + b_{j+1} p^{j+1} + \dots,$$

where $0 \le a_i < p$, $0 \le b_i < p$, $a_j \neq 0$, and either $b_j = 0$ or $b_j = a_j$. If $b_j = 0$, then $n^* \supset m^*$ and $n^* \not\supset m^* - 1$; also $m > 1$, since $m^* \ge 1$. If $b_j \neq 0$, then $b_j = a_j$ and hence $n^* \supset m^* - 1$, $n^* \not\supset m^*$. Thus, in either case

$$f_{p,e}(n - 1, m) + f_{p,e}(n - 1, m - 1) = 1.$$

■

The proof of the next lemma requires a great deal of delicate footwork!

19.9 LEMMA. Assume that $m \geqq 1$, and

$$f_{p,e}(n - 1, m) + f_{p,e}(n - 1, m - 1) > f_{p,e}(n, m).$$

Then condition (A), (B) or (C), below, is true:

(A) $m = 1$, $e \nmid n$.

(B) $e \mid m$, $e \mid (n + 1)$, and $m^* = p^i$ for some $i \geqq 0$.

(C) There exists an integer j satisfying all the following conditions:

(i) $1 \leqq j < m$,

(ii) $e \mid (n + 1 - m - j)$,

(iii) $(n + 1 - m - j)^* \equiv 0$ modulo $p^{\ell_p(m^*)}$,

(iv) $f_{p,e}(n, j) = f_{p,e}(n, m) = 0$,

(v) $f_{p,e}(n - 1, j) + f_{p,e}(n - 1, j - 1) = f_{p,e}(n - 1, m) + f_{p,e}(n-1,$

Proof: By the last lemma, $f_{p,e}(n, m) = 0$, so we have:

(19.10) If $e \mid m$ or $e \mid (n + 1 - m)$, then $(n + 1)^* \not\supset m^*$.

Assume that conclusions (A) and (B) of the lemma are false.

Step 1 We prove that $m > 1$.

If $m = 1$, then $e \mid n$, since conclusion (A) is false, and $e \neq 1$, since conclusion (B) is false. Therefore, $m^* = 0$ and $(n + 1)^* \supset m^*$, contradicting (19.10).

Step 2 Either (19.11) or (19.12) is true:

(19.11) $e \mid m$, $e \mid (n + 1)$, and either $n^* \supset m^*$ or $n^* \supset (m - 1)^*$.

(19.12) $n^* \supset m^*$, $e \nmid n$, and either $e \mid (m - 1)$ or $e \mid (n - m)$.

Suppose that $f_{p,e}(n - 1, m) = 1$. Then $n^* \supset m^*$ and either $e \mid m$ or $e \mid (n - m)$. If $e \mid m$, then $(n + 1)^* \neq n^*$ by (19.10), so $e \mid (n + 1)$, and conclusion (19.11) holds. If $e \nmid m$, then $e \mid (n - m)$, and conclusion (19.12) follows.

Suppose that $f_{p,e}(n - 1, m - 1) = 1$. Then, since $m - 1 \neq 0$, $n^* \supset (m - 1)^*$ and either $e \mid (m - 1)$ or $e \mid (n + 1 - m)$. If $e \mid (n + 1 - m)$, then $(n + 1)^* \not\supset m^*$, by (19.10), while $n^* \supset (m - 1)^*$; thus either $e \mid m$ or $e \mid (n + 1)$. If $e \mid (n + 1 - m)$, we therefore get conclusion (19.11). If $e \nmid (n + 1 - m)$, then $e \mid (m - 1)$ and $e \nmid n$, so $e \neq 1$ and $(m - 1)^* = m^*$; we now have conclusion (19.12).

Step 3 Assume that $e \mid m$, $e \mid (n + 1)$, and either $n^* \supset m^*$ or $n^* \supset (m - 1)^*$.

We have here that $(m - 1)^* = m^* - 1$, $(n + 1)^* = n^* + 1$, and $n^* + 1 \not\supset m^*$ (by 19.10); in particular, $m^* \neq 0$. Because conclusion (B) is false, $m^* \neq 1$.

Since $n^* \supset m^*$ or $n^* \supset m^* - 1$, there exists a unique integer k such that

$$0 \leq k < m^* \quad \text{and} \quad n^* \equiv m^* + k - 1 \text{ modulo } p^{\ell_p(m^*)}$$

But $k \neq 0$, for otherwise $n^* + 1 \supset m^*$.

Let $j = ke$. Then $j^* = k$, $(j - 1)^* = k - 1$, and

$1 \leq j < m$,

$e \mid (n + 1 - m - j)$, and

$(n + 1 - m - j)^* = n^* + 1 - m^* - k \equiv 0$ modulo $p^{\ell_p(m^*)}$.

Also, $n^* + 1 \not\supset j^*$, since $n^* + 1 \not\supset m$, so $f_{p,e}(n, j) = 0$.

Using the fact that e divides j, m, $(n + 1 - j)$, and $(n + 1 - m)$, and the congruence $n^* \equiv m^* + j^* - 1$ modulo $p^{\ell_p(m^*)}$, we have:

$$f_{p,e}(n - 1, j) = 1 \Leftrightarrow n^* \supset j^* \Leftrightarrow n^* \supset m^* - 1 \Leftrightarrow f_{p,e}(n - 1, m - 1) = 1.$$

$$f_{p,e}(n - 1, j - 1) = 1 \Leftrightarrow n^* \supset j^* - 1 \Leftrightarrow n^* \supset m^* \Leftrightarrow f_{p,e}(n - 1, m) = 1.$$

This proves that (19.11) implies conclusion (C).

Step 4 Assume that $n^* \supset m^*$, $e \nmid n$, and either $e \mid (m - 1)$ or $e \mid (n - m)$.

Since $n^* \supset m^*$, either $m^* = 0$ (in which case, $e \nmid (m - 1)$, and we let $k = 0$), or there exists a unique integer k such that

$$0 \leq k < m^* \quad \text{and} \quad n^* \equiv m^* + k \text{ modulo } p^{\ell_p(m^*)}.$$

Now, $e \nmid n$ (in particular, $e \neq 1$), so we may write $n = n^*e + t$, with $1 \leq t < e$. Let

$j = ke + t$ if $e \mid (m - 1)$

$j = ke + 1$ if $e \nmid (m - 1)$; in this case $t \neq 1$.

Then $1 \leq j < m$ and $n^* \supset j^* = (j - 1)^* = k$. Also,

$n + 1 - m - j = n^*e - m^*e - ke$, if $e \mid (m - 1)$,

$n + 1 - m - j = n - m - ke$, if $e \nmid (m - 1)$.

But if $e \nmid (m - 1)$, then $e \mid (n - m)$, and therefore, in either case,

$e \mid (n + 1 - m - j)$ and $(n + 1 - m - j)^* = n^* - m^* - k \equiv 0$ modulo $p^{\ell_p(m^*)}$.

Because $e \nmid j$ and $e \nmid (n + 1 - j)$, we have $f_{p,e}(n, j) = 0$.

If $e \mid (m - 1)$, then $n^* \supset (m - 1)^* = m^*$, $e \mid (n - j)$, $e \nmid (n + 1 - j)$, and $e \mid (j - 1) \Leftrightarrow e \mid (n - m)$. Thus,

$$f_{p,e}(n - 1\ , j) = f_{p,e}(n - 1, m - 1) = 1, \text{ and}$$

$$f_{p,e}(n - 1, j - 1) = 1 \Leftrightarrow f_{p,e}(n - 1, m) = 1.$$

If $e \nmid (m - 1)$, then $e \mid (n - m)$, $e \mid (j - 1)$, and none of j, $(n - j)$, $(n + 1 - j)$, $(n + 1 - m)$ is divisible by e, so

$$f_{p,e}(n - 1, j) = f_{p,e}(n - 1, m - 1) = 0,$$

$$f_{p,e}(n - 1, j - 1) = f_{p,e}(n - 1, m) = 1.$$

This proves that (19.12) implies conclusion (C), and completes the proof of the lemma. ■

The converse of the next lemma is also true, but we do not need it.

19.13 LEMMA. _Assume that $m \geq 1$, and $f_{p,e}(n, m) = 1$. Then there exist integers $j_0, j_1, \ldots, j_{s+1}$ such that_

$$m = j_0 > j_1 > \ldots > j_s > j_{s+1} = 0,$$

and for each i with $1 \leq i \leq s + 1$,

$$e \text{ divides } (n + 1 - j_{i-1} - j_i), \text{ and}$$

$$(n + 1 - j_{i-1} - j_i)^* \equiv 0 \text{ modulo } p^{\ell_p(j^*_{i-1})}.$$

Proof: If $m^* = 0$, then $\ell_p(m^*) = 0$, and because $f_{p,e}(n, m) = 1$ and $m \geq 1$, $e \nmid m$ but $e \mid (n + 1 - m)$. Thus, we may take $m = j_0 > j_1 = 0$. Assume, therefore, that $m^* > 0$. We begin by defining certain integers $k_0, k_1, \ldots, k_{t+1}$ which depend upon m^*.

Let $m^* = k_0$, and suppose that we have defined

$$m^* = k_0 > k_1 > \ldots > k_r > 0 \qquad (r \geq 0)$$

such that

$$(n+1)^* \equiv k_{i-1} + k_i \text{ modulo } p^{\ell_p(k_{i-1})} \quad \text{for } 1 \le i \le r.$$

Then there exists a unique integer k_{r+1} such that

$$0 \le k_{r+1} < k_r \quad \text{and} \quad (n+1)^* \equiv k_r + k_{r+1} \text{ modulo } p^{\ell_p(k_r)}.$$

Thus, there exist integers $t \ge 0$, and $k_0, k_1, \ldots, k_{t+1}$ such that

$$m^* = k_0 > k_1 > \ldots > k_t > k_{t+1} = 0$$

and

$$(n+1)^* \equiv k_{i-1} + k_i \text{ modulo } p^{\ell_p(k_{i-1})} \quad \text{for } 1 \le i \le t+1.$$

Define x by $n+1 = (n+1)^*e + x$, so $0 \le x < e$.

If $e \mid m$, let $j_i = k_i e + x$, if i is odd,

$j_i = k_i e$, if i is even.

If $e \mid (n+1-m)$, then $m = m^*e + x$; let

$j_i = k_i e$, if i is odd,

$j_i = k_i e + x$, if i is even.

Further, let $j_{t+2} = k_{t+2} = 0$. Then $j_i^* = k_i$ $(1 \le i \le t+2)$, and

$$m = j_0 > j_1 > \ldots > j_{t+1} \ge j_{t+2} = 0.$$

Put $s = t$ if $j_{t+1} = 0$, $s = t+1$ if $j_{t+1} > 0$. Then

$$n + 1 - j_{i-1} - j_i = (n+1)^*e - k_{i-1}e - k_i e \quad (1 \le i \le s+1)$$

Therefore, e divides $(n+1-j_{i-1}-j_i)$, and

$$(n+1-j_{i-1}-j_i)^* = (n+1)^* - k_{i-1} - k_i \equiv 0 \text{ mod } p^{\ell_p(j_{i-1}^*)}. \quad ■$$

Theorem 19.6 shows that, when $n - m \ge m > 0$, the trivial $\bar{K}G_n$-module is a submodule of $S_{(n-m,m)}$ if and only if

$$e \mid (n + 1 - m) \quad \text{and} \quad (n + 1)^* \equiv m^* \text{ modulo } p^{\ell_p(m^*)} .$$

These conditions certainly imply that $f_{p,e}(n, m) = 1$. We shall apply the last lemma by showing that if $f_{p,e}(n, m) = 1$, then there is a sequence of modules $S_{(n-j,j)}$, starting at $j = 0$ and ending at $j = m$, each of which has a trivial composition factor.

20 THE COMPOSITION FACTORS OF $S_{(n-m,m)}$

We shall now determine the composition factors (together with composition multiplicities) of S_λ when λ has just two non-zero parts. In view of the form of the Kernel Intersection Theorem, where all the homomorphi involve only changes in two adjacent parts of a partition, it is likely that the result for two-part partitions will shed some light on the general problem of determining the composition factors of S_λ.

Remember that $S_{(n)} = D_{(n)}$ = the trivial $\bar{K}G_n$-module, and $S_{(n-m,m)} \cong S_{(m,n-m)}$, so we may assume that $n - m \geq m$. Since $S_{(n-m,m)}$ is irreducible when K has characteristic zero, we consider only the case where the characteristic of K is a prime p, and $p \nmid q$. As usual, we view $M_{(n-m,m)}$ as the vector space over $\bar{K}$ spanned by the m-dimensional subspaces of V.

We first prove a necessary and sufficient criterion for $S_{(n-m,m)}$ to be irreducible.

20.1 DEFINITION. Let $\mu = (\mu_1, \mu_2, \ldots)$ be a partition of n.

(i) The <u>hook length</u> h_{ij} of the (i, j) mode in the diagram $[\mu]$ is defined by $h_{ij} = \mu_i + \mu'_j + 1 - i - j$ (μ'_j being the length of column j of $[\mu]$)

(ii) The <u>hook graph</u> for μ is obtained by replacing each (i, j) node in $[\mu]$ by h_{ij}.

(iii) Given q and p, let e be the least positive integer such that p divides $(q^e - 1)$, as in Definition 19.4. Let $[\mu]_{p,e}$ be the array of symbols obtained by replacing each (i, j) node in $[\mu]$ by

$\nu_p(h_{ij})$, if e divides h_{ij},

∞, if e does not divide h_{ij}.

20.2 EXAMPLE. If $\mu = (4, 3, 1)$, then the hook graph for μ is

6 4 3 1
4 2 1
1

If $p = 3$ and $q = 2$, then $e = 2$, and

$$[\mu]_{3,2} = \begin{matrix} 1 & 0 & \infty & \infty \\ 0 & 0 & \infty & \\ \infty & & & \end{matrix}$$

If $p = 13$ and $q = 3$, then $e = 3$, and

$$[\mu]_{13,3} = \begin{matrix} 0 & \infty & 0 & \infty \\ \infty & \infty & \infty & \\ \infty & & & \end{matrix}$$

We emphasize that we revert to taking q to be the integer such that $G_n = GL_n(q)$, K is a field of prime characteristic p, with $p \nmid q$, and e is the exponent of q modulo p. $S_{(n-m,m)}$ is a $\bar{K}G_n$-module (or a KG_n-module; see the remarks following Corollary 11.14), and $n - m \geq m$.

20.3 THEOREM. $\underline{S_{(n-m,m)}}$ _is irreducible if and only if no column of_ $\underline{[n - m, m]_{p,e}}$ _contains two different symbols_.

Proof: $S_{(n-m,m)}$ is reducible if and only if it has a composition factor isomorphic to some $D_{(n-j,j)}$ with $0 \leq j < m$, by Corollary 16.3. Therefore, by Theorem 17.1, $S_{(n-m,m)}$ is reducible if and only if $m \geq 1$ and either $S_{(n-m-1,m-1)}$ is reducible or every composition factor of $S_{(n-m,m)}$, besides $D_{(n-m,m)}$, is isomorphic to $D_{(n)}$. Therefore, $S_{(n-m,m)}$ is reducible, by induction, if some column of $[n - m, m]_{p,e}$ after the first column contains two different symbols. Let us assume, therefore, that $m \geq 1$ and no column of $[n - m, m]_{p,e}$, after the first, contains two different symbols. Then, since $D_{(n-m,m)}$ occurs only as a top composition factor of $S_{(n-m,m)}$, $S_{(n-m,m)}$ is reducible if and only if it contains a trivial KG_n-submodule. But this is equivalent to

e divides $(n + 1 - m)$, and

$(n + 1 - m)^* \equiv 0$ modulo $p^{\ell_p(m^*)}$,

by Theorem 19.6.

If $m < e$, then the second of these conditions holds, trivially, and the entry in the (2, 1) place of $[n - m, m]_{p,e}$ is ∞. The entry in the (1, 1 place is not ∞ if and only if e divides $n + 1 - m$, and this proves the theor in the case where $m < e$.

Assume henceforth that $m \geq e$. Since no column of $[n - m, m]_{p,e}$ after the first column contains two different symbols,

$$e \text{ divides } m \Leftrightarrow e \text{ divides } (n + 1 - m), \text{ and}$$

$$(20.4) \qquad \nu_p((m - 1)^* - i) = \nu_p((n - m)^* - i) \quad \text{for } i \leq (m - 1)^* - 1.$$

If both symbols in the first column of $[n - m, m]_{p,e}$ are ∞, then $e \nmid (n + 1$ so $S_{(n-m,m)}$ is irreducible. Let us therefore assume that $e \mid (n + 1 - m)$ an $e \mid m$. Then

$$(n + 1 - m)^* = (n - m)^* + 1 \quad \text{and} \quad m^* = (m - 1)^* + 1.$$

Let $k = \ell_p(m^*)$. Then $p^{k-1} - 1 \leq (m - 1)^* < p^k - 1$, and we deduce from conditions (20.4) that

$$(n - m)^* = b_0 + b_1 p + \ldots + b_{k-2}p^{k-2} + b_{k-1}p^{k-1} + \ldots \qquad (0 \leq b_i < p)$$

$$(m - 1)^* = b_0 + b_1 p + \ldots + b_{k-2}p^{k-2} + a_{k-1}p^{k-1}, \text{ where}$$

either $0 < a_{k-1} \leq b_{k-1}$,

or $\quad b_0 = b_1 = \ldots = b_{k-2} = p - 1 \quad$ and $\quad a_{k-1} = 0$.

But now

$$\nu_p(m^*) \neq \nu_p((n + 1 - m)^*) \Leftrightarrow b_0 = \ldots = b_{k-2} = b_{k-1} = p - 1$$

$$\Leftrightarrow (n + 1 - m)^* \equiv 0 \text{ modulo } p^k.$$

That is, the two numbers in the first column of $[n - m, m]_{p,e}$ are unequal if and only if $S_{(n - m, m)}$ is reducible. ∎

It has been proved (James $[J_5]$ and James and Murphy [JM]) that in ymmetric group theory, the Specht module for the partition μ (μ p-regular) s irreducible if and only if no column of $[\mu]_{p,1}$ contains two different umbers. Since e = 1 when "q = 1", this lends credence to the following onjecture, which is supported by the last theorem.

0.5 CONJECTURE. *S_μ is irreducible if and only if no column of $[\mu]_{p,e}$ ontains two different symbols.*

In view of Corollary 16.3, which tells us that every composition actor of $S_{(n-m,m)}$ has the form $D_{(n-j,j)}$ with $0 \le j \le m$, the main theorem of his chapter determines all the composition factors of $S_{(n-m,m)}$:

0.6 THEOREM. *Let $S_{(n-m,m)}$ be defined over a field of prime characteristic , where p does not divide q. Then the composition multiplicity of $D_{(n-j,j)}$ n $S_{(n-m,m)}$ for $0 \le j \le m$ is $f_{p,e}(n-2j, m-j)$, given by Definition 19.7.*

Theorem 17.1 reduces this theorem to a matter of proving:

0.7 THEOREM. *The composition multiplicity of $D_{(n)}$ in $S_{(n-m,m)}$ is $f_{p,e}(n, m)$.*

roof: *Step 1*. Assume that $f_{p,e}(n, m) = 1$. We prove that the composition ultiplicity of $D_{(n)}$ in $S_{(n-m,m)}$ is at least 1.

Suppose that $0 \le j < k \le n - k$. Let θ be the $\bar{K}G_n$-homomorphism from $_{(n-j,j)}$ into $M_{(n-k,k)}$ which sends each j-dimensional subspace V_1 of V to the um of the k-dimensional subspaces which intersect V_1 in zero.

We claim that Ker $\theta \not\supseteq S_{(n-j,j)}$. The simplest way to see this is in erms of the generator of $S_{(n-j,j)}$ described in Section 18, so let V* be a j-dimensional subspace of V, spanned by $f_1, \ldots, f_j, g_1, \ldots, g_j$, say. Then

$$\xi = \sum_W (\text{sign } W) W,$$

ummed over the j-dimensional isotropic subspaces W of V*, belongs to $S_{(n-j,j)}$.

(See Definition 18.11 and Theorem 18.13). Let $\langle h_1, \ldots, h_{n-2j}\rangle$ be a comple

to V* in V. Then the subspace

$$V_2 = \langle f_1, \ldots, f_j, h_1, \ldots, h_{k-j}\rangle$$

occurs in $\xi\theta$ with coefficient equal to

$$\Sigma\{\text{sign } W \mid W \text{ is a } j\text{-dimensional isotropic subspace of } V^* \text{ and } W \cap \langle f_1, \ldots, f_j, h_1, \ldots, h_{k-j}\rangle = 0\}.$$

But $\langle f_1, \ldots, f_j\rangle$ is isotropic, and there exist $q^{j(j-1)/2}$ j-dimensional isotropic subspaces of V* which intersect this one in zero, by Corollary 18 all these $q^{j(j-1)/2}$ subspaces have the same sign. Therefore,

$$\langle V_2, \xi\theta\rangle_{(n-k,k)} = \pm q^{j(j-1)/2} \neq 0.$$

In particular, $\xi\theta \neq 0$, and $\text{Ker } \theta \not\supseteq S_{(n-j,j)}$.

We now know that $\text{Ker } \theta \subseteq S^{\perp}_{(n-j,j)}$, by the Submodule Theorem. Since $M_{(n-j,j)}/S^{\perp}_{(n-j,j)}$ is isomorphic to the dual of $S_{(n-j,j)}$,

(20.8) Every composition factor of $S_{(n-j,j)}$ is isomorphic to a composition factor of $M_{(n-j,j)}\theta$.

Next, we consider $M_{(n-j,j)}\theta$. Let $0 \leq b < k$, and V_1, V_3 be subspaces of V with $\dim V_1 = j$, $\dim V_3 = b$, and $V_1 \cap V_3 = 0$. Then, by Theorem 3.1, t number of k-dimensional subspaces of V which both contain V_3 and intersect in zero is

$$q^{j(k-b)} \begin{bmatrix} n-j-b \\ k-b \end{bmatrix}.$$

Therefore,

$$M_{(n-j,j)}\theta \subseteq \text{Ker } \psi_{1,b} \Leftrightarrow p \text{ divides } \begin{bmatrix} n-j-b \\ k-b \end{bmatrix}.$$

The Kernel Intersection Theorem and Theorem 19.5 now give:

$$M_{(n-j,j)}\theta \subsetneq S_{(n-k,k)}$$

$\Leftrightarrow$ e divides $(n + 1 - j - k)$ and $(n + 1 - j - k)^* \equiv 0$ modulo $p^{\ell_p(k^*)}$.

Combining this with (20.8), we have proved:

(20.9) Every composition factor of $S_{(n-j,j)}$ is isomorphic to a composition factor of $S_{(n-k,k)}$ if conditions (i) - (iii) hold:

(i) $0 \leq j < k \leq n - k$

(ii) e divides $(n + 1 - k - j)$

(iii) $(n + 1 - k - j)^* \equiv 0$ modulo $p^{\ell_p(k^*)}$.

Now assume that $f_{p,e}(n, m) = 1$. Then, by Lemma 19.13, there exist integers,

$$m = j_0 > j_1 > \dots > j_s > j_{s+1} = 0$$

such that, for each i with $1 \leq i \leq s + 1$,

e divides $(n + 1 - j_{i-1} - j_i)$, and

$(n + 1 - j_{i-1} - j_i)^* \equiv 0$ modulo $p^{\ell_p(j^*_{i-1})}$.

Since $D_{(n)} = S_{(n)}$, we now deduce, by repeatedly applying (20.9) with $j = j_i$, $k = j_{i-1}$ for $i = s + 1, s, \dots, 1$ in turn, that $D_{(n)}$ is isomorphic to a composition factor of $S_{(n-m,m)}$.

<u>Step 2</u> The composition multiplicity of $D_{(n)}$ in $S_{(n-m,m)}$ is at most $f_{p,e}(n, m)$.

This is certainly true if $m = 0$, so it is true when $n = 0$ (and when $n = 1$). Therefore, we may assume that $m \geq 1$ and that the result of Step 2 is true for all two-part partitions of $n - 1$.

Let χ_λ denote the p-modular character of $S_{\lambda,\mathbb{Q}}$. If the composition multiplicity of $D_{(n)}$ in $S_{(n-m,m)}$ is x, then $\chi_{(n-m,m)} - x\chi_{(n)}$ is a p-modular character of G_n, and Theorem 16.11 shows that

$$\chi_{(n-m-1,m)} + \chi_{(n-m,m-1)} - x\chi_{(n-1)}$$

is a p-modular character of G_{n-1}. (Here, the first term must be omitted if $n = 2m$; but we have seen that $f_{p,e}(n - 1, m) = 0$ if $n = 2m$, so this makes no odds.) Hence, by induction,

$$f_{p,e}(n - 1, m) + f_{p,e}(n - 1, m - 1) \geq x.$$

Therefore, the result of step 2 is immediate unless

$$f_{p,e}(n - 1, m) + f_{p,e}(n - 1, m - 1) > f_p(n, m),$$

so we assume that this inequality holds.

We aim to prove:

(20.10) There exists j with $1 \leq j \leq m$ such that some composition factor of $S_{(n-m,m)}$ is isomorphic to $D_{(n-j,j)}$, and when we regard $D_{(n-j,j)}$ as a KG_{n-1}-module, it has $D_{(n-1)}$ as a composition factor with multiplicity at least

$$f_{p,e}(n - 1, m) + f_{p,e}(n - 1, m - 1).$$

At least one of conclusions (A), (B), (C) of Lemma 19.9 is true. We examine each possibility in turn.

If $m = 1$ and $e \nmid n$, then the two symbols in the first column of $[n - m, m]_{p,e}$ are ∞. Therefore $S_{(n - m, m)}$ is irreducible, by Theorem 20.3. Here, we may take $j = m$, when (20.10) follows from the Branching Theorem.

Assume next that $e \mid m$, $e \mid (n + 1)$, and $m^* = p^k$ where $k \geq 0$. We shal show that $S_{(n-m,m)}$ is irreducible in this case, too. Certainly,

$$e \mid (m - i) \Leftrightarrow e \mid (n + 1 - m - i) \quad \text{for } 0 \leq i \leq m - 1.$$

Also, $(m - 1)^* = m^* - 1$, $n^* = (n + 1)^* - 1$, and n^* contains either m^* or m^* - to base p, and $n^* + 1$ does not contain m^* to base p (by Lemma 19.8). Therefo

$$n^* = (p - 1) + \ldots + (p - 1)p^{k-1} + b_k p^k + ap^{k+1}$$

with $0 < b_k < p$, $0 \leq a$. Thus,

$$(n + 1 - m)^* = n^* + 1 - m^* = b_k p^k + ap^{k+1},$$

hence, for $0 \le i \le m^* - 1$,

$$\nu_p(m^* - i) = \nu_p((n + 1 - m)^* - i).$$

refore, Theorem 20.3 proves that $S_{(n-m,m)}$ is irreducible. Again, we may

$j = m$ in (20.10).

Finally, assume that there exists an integer j satisfying the conditions

ed in conclusion (C) of Lemma 19.9. Applying (20.9), we have:

(i) $1 \le j < m$,

(ii) $f_{p,e}(n, j) = f_{p,e}(n, m) = 0$

(iii) $f_{p,e}(n - 1, j) + f_{p,e}(n - 1, j - 1) = f_{p,e}(n - 1, m) + f_{p,e}(n - 1, m - 1)$

(iv) Every composition factor of $S_{(n-j,j)}$ is isomorphic to a composition

or of $S_{(n-m,m)}$.

Since $1 \le j < m$ and

$$f_{p,e}(n - 1, j) + f_{p,e}(n - 1, j - 1) > f_{p,e}(n, j),$$

educe, by induction in (20.10), that there exists k with $1 \le k \le j$ such

some composition factor of $S_{(n-j,j)}$ is isomorphic to $D_{(n-k,k)}$, and when

egard $D_{(n-k,k)}$ as a KG_{n-1}-module, it has $D_{(n-1)}$ as a composition factor

multiplicity at least

$$f_{p,e}(n - 1, j) + f_{p,e}(n - 1, j - 1) = f_{p,e}(n - 1, m) + f_{p,e}(n - 1, m - 1).$$

some composition factor of $S_{(n-m,m)}$ is isomorphic to $D_{(n-k,k)}$, and so the

f of (20.10) is complete.

Let $\phi_{(n-j,j)}$ be the p-modular character of $D_{(n-j,j)}$ in (20.10), and

n denote by x the composition multiplicity of $D_{(n)}$ in $S_{(n-m,m)}$. Then

$_{m,m)} - \phi_{(n-j,j)} - x\chi_{(n)}$ is a p-modular character of G_n, and

$$\chi_{(n-m-1,m)} + \chi_{(n-m,m-1)} - (x + y)\chi_{(n-1)}$$

is a p-modular character of G_{n-1}, where, by (20.10),

$$y \geq f_{p,e}(n - 1, m) + f_{p,e}(n - 1, m - 1).$$

Therefore, $x = 0 = f_{p,e}(n, m)$, and this completes the proof of Step 2.

Putting together the results of Steps 1 and 2, Theorem 20.7 is prov

Theorem 20.6 tells us that the composition multiplicity of $D_{(n-j,j)}$ in $S_{(n-m,m)}$ is 1 if

$(n + 1 - 2j)^*$ contains $(m - j)^*$ to base p, and

either $e \mid (m - j)$ or $e \mid (n + 1 - m - j)$,

and is 0, otherwise. The corresponding theorem for the symmetric group $\mathfrak{S}_n$ (James $[J_2, J_3]$) states that the p-modular irreducible representation indexed by $(n - j, j)$ is a composition factor of the Specht module for $(n - m, m)$ with composition multiplicity 1 if

$(n + 1 - 2j)$ contains $(m - j)$ to base p,

and with composition multiplicity 0, otherwise. But if $q = 1$ then $e = 1$, so the symmetric group theorem is just that obtained by putting $q = 1$ in t theorem for the general linear group.

We could hardly conclude with a more striking illustration of our main tenet, namely that the representation theory of the symmetric group h a "q-analogue" in the general linear group theory. Everything we have don easily reduces to a result for the symmetric group when q is replaced by 1 but it is far from clear how to reverse the process at each stage; conside the main theorem of this chapter, for example - it would be extremely hard to guess the correct q-analogue of the statement that $(n + 1 - 2j)$ contains $(m - j)$ to base p. We believe therefore, that the theory of unipotent representation of $GL_n(q)$ is a deep and significant topic which will fully repay further research.

CKNOWLEDGEMENTS

I am indebted to Jorn Olsson for his suggestion in February 1980 that should undertake an investigation of unipotent representations of the eneral linear groups. My gratitude is also due to Hanafi Farahat for many seful conversations during a visit to the University of Calgary in the ummer of 1982, when the final parts of the research for this essay were ompleted.

REFERENCES

A G. E. ANDREWS, "The Theory of Partitions", Encyclopedia of Math. Appl., Vol. 2, Addison-Wesley, 1976

B R. BRAUER, On a conjecture by Nakayama, Trans. Roy. Soc. Canada Sect III (3) 41 (1947), 11-19

C R. W. CARTER, "Simple Groups of Lie Type", Wiley, London, 1972

CL R. W. CARTER and G. LUSZTIG, On the modular representations of the general linear and symmetric groups, Math. Zeit. 136 (1974), 193-2

D L. DORNHOFF, "Group Representation Theory, Part A", Dekker, 1971

FS P. FONG and B. SRINIVASAN, The blocks of finite general linear and unitary groups, Inventiones Math. 69 (1982), 109-153

FRT J. S. FRAME, G. de B. ROBINSON, and R. M. THRALL, The hook graphs the symmetric group, Can. J. Math. 6 (1954), 316-324

G_1 J. A. GREEN, The characters of the finite general linear groups, Trans. Amer. Math. Soc. 80 (1955), 402-447

G_2 J. A. GREEN, "Polynomial Representations of GL_n", Lecture Notes in Math., Vol. 830, Springer Verlag, 1980

J_1 G. D. JAMES, The irreducible representations of symmetric groups, Bull. London Math. Soc. 8 (1976), 229-232

J_2 G. D. JAMES, Representations of the symmetric group over the field order 2, J. Algebra 38 (1976), 280-308

J_3 G. D. JAMES, On the decomposition matrices of the symmetric groups J. Algebra 43 (1976), 42-44

J_4 G. D. JAMES, A characteristic-free approach to the representation theory of $\mathfrak{S}_n$, J. Algebra 46 (1977), 430-450

J_5 G. D. JAMES, On a conjecture of Carter concerning irreducible Spec modules, Math. Proc. Camb. Philos. Soc. 83 (1978), 11-17

J_6 G. D. JAMES, "The Representation Theory of the Symmetric Groups", Lecture Notes in Math., Vol. 682, Springer-Verlag, 1978

J_7 G. D. JAMES, The decomposition of tensors over fields of prime characteristic, Math. Zeit. 172 (1980), 161-178

J_8 G. D. JAMES, On the decomposition matrices of the symmetric groups J. Algebra, 71 (1981), 115-122

J_9 G. D. JAMES, Unipotent representations of the finite general linea groups, J. Algebra, 74 (1982), 443-465

J_{10} G. D. JAMES, Trivial source modules for symmetric groups, submitte Math. Zeit.

JK G. D. JAMES and A. KERBER, "The Representation Theory of the Symmetric Group", Encyclopedia of Math. Appl., Vol. 16, Addison-Wesley, 1981

JM G. D. JAMES and G. E. MURPHY, The determinant of the Gram matrix for a Specht module, J. Algebra 59 (1979), 222-235

L S. LANG, "Algebra", Addison-Wesley, 1965

O J. B. OLSSON, On the blocks of GL(n, q)I, Trans. Amer. Math. Soc. 222 (1976), 143-156

P M. H. PEEL, Hook representations of the symmetric groups, Glasgow Math. J. 12 (1971), 136-149

R G. de B. ROBINSON, On a conjecture by Nakayama, Trans. Roy. Soc. Canada. Sct. III (3) 41 (1947), 20-25

S R. STEINBERG, A geometric approach to the representations of the full linear group over a Galois field, Trans. Amer. Math. Soc. 71 (1951), 274-282

T G. P. THOMAS, Further results on Baxter sequences and generalised Schur functions, in "Combinatoire et représentation du groupe symétrique", Lecture Notes in Math., Vol. 579, Springer-Verlag, 1977